Lecture Notes in Mathematics 1772

Editors:
J.-M. Morel, Cachan
F. Takens, Groningen
B. Teissier, Paris

W0055432

Springer
Berlin
Heidelberg
New York
Barcelona
Hong Kong
London
Milan
Paris
Tokyo

F. E. Burstall D. Ferus
K. Leschke F. Pedit U. Pinkall

Conformal Geometry
of Surfaces in S^4
and Quaternions

Springer

Authors

Francis E. Burstall
Dept. of Mathematical Sciences
University of Bath
Claverton Down
Bath BA2 7AY, U.K.
E-mail: f.e.burstall@maths.bath.ac.uk

Dirk Ferus
Katrin Leschke
Technical University of Berlin
MA 8-3
Strasse des 17. Juni 136
10623 Berlin, Germany
E-mail: ferus@math.tu-berlin.de
E-mail: leschke@math.tu-berlin.de

Franz Pedit
Dept. of Mathematics
and Statistics
University of Massachusetts
1542, Lederle
Amherst, MA 01003, U.S.A.
E-mail: franz@gang.umass.edu

Ulrich Pinkall
Technical University of Berlin
MA 8-3
Strasse des 17. Juni 136
10623 Berlin, Germany
E-mail: pinkall@math.tu-berlin.de

Cover figure from D. Ferus, F. Pedit:S^1-equivariant Minimal Tori in S^4 and S^1-equivariant Willmore Tori in S^3. Math. Z. 204, 269-282 (1990)

Cataloging-in-Publication Data applied for.

Die Deutsche Bibliothek - CIP-Einheitsaufnahme

Conformal geometry of surfaces in S4 and quaternions / F. E. Burstall ....
- Berlin ; Heidelberg ; New York ; Barcelona ; Hong Kong ; London ; Milan ; Paris ; Tokyo :
- Springer, 2002
    (Lecture notes in mathematics ; 1772)
    ISBN 3-540-43008-3

Mathematics Subject Classification (2000): 53C42, 53A30

ISSN 0075-8434
ISBN 3-540-43008-3 Springer-Verlag Berlin Heidelberg New York

Springer-Verlag Berlin Heidelberg New York a member of BertelsmannSpringer
Science + Business Media GmbH

http://www.springer.de

© Springer-Verlag Berlin Heidelberg 2002

Typesetting: Camera-ready TEX output by the authors

SPIN: 10856631   41/3142-543210 - Printed on acid-free paper

Preface

This is the first comprehensive introduction to the authors' recent attempts toward a better understanding of the global concepts behind spinor representations of surfaces in 3-space. The important new aspect is a quaternionic-valued function theory, whose "meromorphic functions" are conformal maps into $\mathbb{H}$, which extends the classical complex function theory on Riemann surfaces. The first results along these lines were presented at the ICM 98 in Berlin [10], and a detailed exposition will appear in [4]. Basic constructions of complex Riemann surface theory, such as holomorphic line bundles, holomorphic curves in projective space, Kodaira embedding, and Riemann-Roch, carry over to the quaternionic setting. Additionally, an important new invariant of the quaternionic holomorphic theory is the Willmore energy. For quaternionic holomorphic curves in $\mathbb{H}P^1$ this energy is the classical Willmore energy of conformal surfaces.

The present lecture note is based on a course given by Dirk Ferus at the Summer School on Differential Geometry at Coimbra in September, 1999, [3]. It centers on Willmore surfaces in the conformal 4-sphere $\mathbb{H}P^1$. The first three sections introduce linear algebra over the quaternions and the quaternionic projective line as a model for the conformal 4-sphere. Conformal surfaces $f : M \to \mathbb{H}P^1$ are identified with the pull-back of the tautological bundle. They are treated as quaternionic line subbundles of the trivial bundle $M \times \mathbb{H}^2$. A central object, explained in section 5, is the mean curvature sphere (or conformal Gauss map) of such a surface, which is a complex structure on $M \times \mathbb{H}^2$. It leads to the definition of the Willmore energy, the critical points of which are called Willmore surfaces. In section 7 we identify the new notions of our quaternionic theory with notions in classical submanifold theory. The rest of the paper is devoted to applications: We classify super-conformal immersions as twistor projections from $\mathbb{C}P^3$ in the sense of Penrose, we construct Bäcklund transformations for Willmore surfaces in $\mathbb{H}P^1$, we set up a duality between Willmore surfaces in S^3 and certain minimal surfaces in hyperbolic 3-space, and we give a new proof of the classification of Willmore 2-spheres in the 4-sphere, see Ejiri [2], Musso [9] and Montiel [8]. Finally we explain a close similarity between the theory of constant mean curvature spheres in $\mathbb{R}^3$ and that of Willmore surfaces in $\mathbb{H}P^1$, and use it to construct Darboux transforms for the latter.

Bath/Berlin,
August 2001

Francis Burstall, Dirk Ferus, Katrin Leschke,
Franz Pedit, Ulrich Pinkall

Table of Contents

1 Quaternions

1.1 The Quaternions

The Hamiltonian quaternions $\mathbb{H}$ are the unitary $\mathbb{R}$-algebra generated by the symbols i, j, k with the relations

$$i^2 = j^2 = k^2 = -1,$$
$$ij = -ji = k, \quad jk = -kj = i, \quad ki = -ik = j.$$

The multiplication is associative but obviously not commutative, and each non-zero element has a multiplicative inverse: We have a skew-field, and a 4-dimensional division algebra over the reals. Frobenius showed in 1877 that $\mathbb{R}, \mathbb{C}$ and $\mathbb{H}$ are in fact the only finite-dimensional $\mathbb{R}$-algebras that are associative and have no zero-divisors. For the element

$$a = a_0 + a_1 i + a_2 j + a_3 k, \quad a_l \in \mathbb{R}, \tag{1.1}$$

we define

$$\bar{a} := a_0 - a_1 i - a_2 j - a_3 k,$$
$$\operatorname{Re} a := a_0,$$
$$\operatorname{Im} a := a_1 i + a_2 j + a_3 k.$$

Note that, in contrast with the complex numbers, $\operatorname{Im} a$ is not a real number, and that conjugation obeys

$$\overline{ab} = \bar{b}\,\bar{a}.$$

We shall identify the real vector space $\mathbb{H}$ in the obvious way with $\mathbb{R}^4$, and the subspace of purely imaginary quaternions with $\mathbb{R}^3$:

$$\mathbb{R}^3 = \operatorname{Im}\mathbb{H}.$$

The reals are identified with $\mathbb{R}1$. The embedding of the complex numbers $\mathbb{C}$ is less canonical. The quaternions i, j, k equally qualify for the complex imaginary unit, and in fact any purely imaginary quaternion of square -1 would do the job. From now on, however, we shall usually use the subfield $\mathbb{C} \subset \mathbb{H}$ generated by $1, i$.

Occasionally we shall need the Euclidean inner product on $\mathbb{R}^4$ which can be written as

$$< a, b >_{\mathbb{R}} = \mathrm{Re}(\bar{a}b) = \mathrm{Re}(a\bar{b}) = \frac{1}{2}(\bar{a}b + \bar{b}a).$$

We define

$$|a| := \sqrt{< a, a >_{\mathbb{R}}} = \sqrt{a\bar{a}}.$$

Then

$$|ab| = |a|\,|b|. \tag{1.2}$$

A closer study of the quaternionic multiplication displays nice geometric aspects.

We first mention that the quaternion multiplication incorporates both the usual vector and scalar products on $\mathbb{R}^3$. In fact, using the representation (1.1) one finds for $a, b \in \mathrm{Im}\,\mathbb{H} = \mathbb{R}^3$

$$ab = a \times b - < a, b >_{\mathbb{R}}. \tag{1.3}$$

As a consequence we state

Lemma 1. *For $a, b \in \mathbb{H}$ we have*

1. *$ab = ba$ if and only if $\mathrm{Im}\,a$ and $\mathrm{Im}\,b$ are linearly dependent over the reals. In particular, the reals are the only quaternions that commute with all others.*
2. *$a^2 = -1$ if and only if $|a| = 1$ and $a = \mathrm{Im}\,a$. Note that the set of all such a is the usual two-sphere*

$$S^2 \subset \mathbb{R}^3 = \mathrm{Im}\,\mathbb{H}.$$

Proof. Write $a = a_0 + a', b = b_0 + b'$, where the prime denotes the imaginary part. Then

$$ab = a_0 b_0 + a_0 b' + a' b_0 + a' b'$$
$$= a_0 b_0 + a_0 b' + a' b_0 + a' \times b' - < a', b' >_{\mathbb{R}}.$$

All these products, except for the cross-product, are commutative, and (1) follows. From the same formula with $a = b$ we obtain $\mathrm{Im}\,a^2 = 2a_0 a'$. This vanishes if and only if a is real or purely imaginary. Together with (1.2) we obtain (2).

1.2 The Group S^3

The set of unit quaternions

$$S^3 := \{\mu \in \mathbb{H} \,|\, |\mu|^2 = 1\}$$

i.e. the 3-sphere in $\mathbb{H} = \mathbb{R}^4$, forms a group under multiplication. We can also interpret it as the group of linear maps $x \mapsto \mu x$ of $\mathbb{H}$ preserving the hermitian inner product

$$< a, b >:= \bar{a}b.$$

This group is called the symplectic group $Sp(1)$.

We now consider the action of S^3 on $\mathbb{H}$ given by

$$S^3 \times \mathbb{H} \to \mathbb{H}, \quad (\mu, a) \mapsto \mu a \mu^{-1}.$$

By (1.2) this action preserves the norm on $\mathbb{H} = \mathbb{R}^4$ and, hence, the Euclidean scalar product. It obviously stabilizes $\mathbb{R} \subset \mathbb{H}$ and, therefore, its orthogonal complement $\mathbb{R}^3 = \operatorname{Im} \mathbb{H}$. We get a map, in fact a representation,

$$\pi : S^3 \to SO(3), \mu \mapsto \mu \ldots \mu^{-1}|_{\operatorname{Im} \mathbb{H}}.$$

Let us compute the differential of π. For $\mu \in S^3$ and $v \in T_\mu S^3 = (\mathbb{R}\mu)^\perp$, we get

$$d_\mu \pi(v)(a) = va\mu^{-1} - \mu a \mu^{-1} v \mu^{-1} = \mu(\mu^{-1}va - a\mu^{-1}v)\mu^{-1}.$$

Now $\mu^{-1}v$ commutes with all $a \in \operatorname{Im} \mathbb{H}$ if and only if $v = r\mu$ for some real r. But then $v = 0$, because $v \perp \mu$. Hence π is a local diffeomorphism of S^3 onto the 3-dimensional manifold $SO(3)$ of orientation preserving orthogonal transformations of $\mathbb{R}^3$. Since S^3 is compact and $SO(3)$ is connected, this is a covering. And since $\mu a \mu^{-1} = a$ for all $a \in \operatorname{Im} \mathbb{H}$ if and only if $\mu \in \mathbb{R}$, i.e. if and only if $\mu = \pm 1$, this covering is 2:1. It is obvious that antipodal points of S^3 are mapped onto the same orthogonal transformation, and therefore we see that

$$SO(3) \cong S^3/\{\mu \sim -\mu\} = \mathbb{R}P^3.$$

We have now displayed the group of unit quaternions as the universal covering of $SO(3)$. This group is also called the spin group:

$$S^3 = Sp(1) = Spin(3).$$

If we identify $\mathbb{H} = \mathbb{C} \oplus \mathbb{C}j = \mathbb{C}^2$, we can add yet another isomorphism:

$$S^3 \cong SU(2).$$

In fact, let $\mu = \mu_0 + \mu_1 j \in S^3$ with $\mu_0, \mu_1 \in \mathbb{C}$. Then for $\alpha, \beta \in \mathbb{R}$ we have $j(\alpha + i\beta) = (\alpha - i\beta)j$. Therefore the $\mathbb{C}$-linear map $A_\mu : \mathbb{C}^2 \to \mathbb{C}^2, x \mapsto \mu x$ has the following matrix representation with respect to the basis $1, j$ of $\mathbb{C}^2$:

$$A_\mu 1 = \quad \mu_0 + \mu_1 j = \quad 1\mu_0 + j\bar{\mu}_1$$
$$A_\mu j = -\mu_1 + \mu_0 j = 1(-\mu_1) + j\bar{\mu}_0.$$

Because of $\mu_0 \bar{\mu}_0 + \mu_1 \bar{\mu}_1 = 1$, we have

$$\begin{pmatrix} \mu_0 & \bar{\mu}_1 \\ -\mu_1 & \bar{\mu}_0 \end{pmatrix} \in SU(2).$$

2 Linear Algebra over the Quaternions

2.1 Linear Maps, Complex Quaternionic Vector Spaces

Since we consider vector spaces V over the skew-field of quaternions, there are two options for the multiplication by scalars. We choose quaternion vector spaces to be *right* vector spaces, i.e. vectors are multiplied by quaternions from the right:

$$V \times \mathbb{H} \to V, (v, \lambda) \mapsto v\lambda.$$

The notions of basis, dimension, subspace, and linear map work as in the usual commutative linear algebra. The same is true for the matrix representation of linear maps in finite dimensions. However, there is no reasonable definition for the elementary symmetric functions like trace and determinant: The linear map $A : \mathbb{H} \to \mathbb{H}, x \mapsto ix$, has matrix (i) when using 1 as basis for $\mathbb{H}$, but matrix $(-i)$ when using the basis j.

If $A \in \mathrm{End}(V)$ is an endomorphism, $v \in V$, and $\lambda \in \mathbb{H}$ such that

$$Av = v\lambda,$$

then for any $\mu \in \mathbb{H} \backslash \{0\}$ we find

$$A(v\mu) = (Av)\mu = v\lambda\mu = (v\mu)(\mu^{-1}\lambda\mu).$$

If λ is real then the eigenspace is a quaternionic subspace. Otherwise it is a real – but *not* a quaternionic – vector subspace, and we obtain a whole 2-sphere of "associated eigenvalues" (see Section 1.2). This is related to the fact that multiplication by a quaternion (necessarily from the right) is *not* an $\mathbb{H}$-linear endomorphism of V. In fact, the space of $\mathbb{H}$-linear maps between quaternionic vector spaces is *not* a quaternionic vector space itself.

Any quaternionic vector space V is of course a complex vector space, but this structure depends on choosing an imaginary unit, as mentioned in section 1.1. We shall instead (quite regularly) have an *additional* complex structure on V, acting from the left, and hence commuting with the quaternionic structure. In other words, we consider a fixed $J \in \mathrm{End}(V)$ such that $J^2 = -I$. Then

$$(x + iy)v := vx + (Jv)y.$$

In this case we call (V, J) a *complex quaternionic (bi-)vector space*. If (V, J) and (W, J) are such spaces, then the quaternionic linear maps from V to W split as a direct sum of the real vector spaces of complex linear $(AJ = JA)$ and anti-linear $(AJ = -JA)$ homomorphisms.

$$\text{Hom}(V, W) = \text{Hom}_+(V, W) \oplus \text{Hom}_-(V, W)$$

In fact, $\text{Hom}(V, W)$ and $\text{Hom}_\pm(V, W)$ are *complex* vector space with multiplication given by

$$(x + iy)Av := (Av)x + (JAv)y.$$

The standard example of a quaternionic vector space is $\mathbb{H}^n$. An example of a *complex* quaternionic vector space is $\mathbb{H}^2$ with $J(a, b) := (-b, a)$.

On $V = \mathbb{H}$, any complex structure is simply left-multiplication by some $N \in \mathbb{H}$ with $N^2 = -1$. The following lemma describes a situation that naturally produces such an N, and that will become a standard situation for us. But, before stating that lemma, let us make a simple observation:

Remark 1. On a real 2-dimensional vector space U each complex structure $J \in \text{End}(U)$ induces an orientation $\mathcal{O}$ such that (x, Jx) is positively oriented for any $x \neq 0$. We then call J compatible with $\mathcal{O}$.

Lemma 2 (Fundamental lemma).
 1. *Let $U \subset \mathbb{H}$ be a real subspace of dimension 2. Then there exist $N, R \in \mathbb{H}$ with the following three properties:*

$$N^2 = -1 = R^2, \tag{2.1}$$
$$NU = U = UR, \tag{2.2}$$
$$U = \{x \in \mathbb{H} \,|\, NxR = x\}. \tag{2.3}$$

The pair (N, R) is unique up to sign. If U is oriented, there is only one such pair such that N is compatible with the orientation.
 2. *If U, N and R are as above, and $U \subset \text{Im}\,\mathbb{H}$, then*

$$N = R,$$

and this is a Euclidean unit normal vector of U in $\text{Im}\,\mathbb{H} = \mathbb{R}^3$.
 3. *Given $N, R \in \mathbb{H}$ with $N^2 = -1 = R^2$, the sets*

$$U := \{x \in \mathbb{H} \,|\, NxR = x\}, \quad U^\perp := \{x \in \mathbb{H} \,|\, NxR = -x\}$$

are orthogonal real subspaces of dimension 2.

Definition 1. *Motivated by (2) of the lemma, N and R are called a* left *and* right *normal vector of U, though in general they are not at all* orthogonal *to U in the geometric sense.*

Proof (of the lemma). (1). If $1 \in U$ and if $a \in U$ is a unit vector orthogonal to 1, then $a^2 = -1$. Hence $(N, R) = (a, -a)$ works for U, and the uniqueness, up to sign, follows easily from $N1 \in U$ and $Na \in U$. If U is arbitrary, and $x \in U \backslash \{0\}$ then put $\tilde{U} := x^{-1}U$. Clearly, $1 \in \tilde{U}$. Moreover, (N, R) works for U if and only if $(x^{-1}Nx, R)$ works for $\tilde{U}$.

(2). If $U \subset \operatorname{Im} \mathbb{H} = \mathbb{R}^3$, and u, v is an orthonormal basis of U, then $N = R = u \times v = uv$ satifies the requirements: Use the geometric properties of the cross product.

(3). The above argument shows that $\sigma(x) := NxR$ has ± 1-eigenspaces of real dimension 2. Since σ is orthogonal, so are its eigenspaces.

Example 1. Let $(V, J), (W, J)$ be complex quaternionic vector spaces of dimension 1. Then $\operatorname{Hom}_+(V, W)$ is of real dimension 2. To see this, choose bases v and w, and assume

$$ Jv = vR, \quad Jw = wN. $$

Then $N^2 = -1 = R^2$. Now $F \in \operatorname{Hom}(V, W)$ is given by $F(v) = wa$, and

$$ FJ = JF \iff FJv = JFv $$
$$ \iff waR = J(wa) = (Jw)a = wNa \iff aR = Na. $$

But the set of all such a is of real dimension 2, by the last part of the lemma. The same result holds for $\operatorname{Hom}_-(V, W)$. As stated earlier, $\operatorname{Hom}_\pm(V, W)$ are complex vector spaces, and therefore (non-canonically) isomorphic with $\mathbb{C}$.

2.2 Conformal Maps

A linear map $F : V \to W$ between Euclidean vector spaces is called *conformal* if there exists a positive λ such that

$$ < Fx, Fy > = \lambda < x, y > $$

for all $x, y \in V$. This is equivalent to the fact that F maps a normalized orthogonal basis of V into a normalized orthogonal basis of $F(V) \subset W$. Here "normalized" means that all vectors have the same length, possibly $\neq 1$.

If $V = W = \mathbb{R}^2 = \mathbb{C}$, and $J : \mathbb{C} \to \mathbb{C}$ denotes multiplication by the imaginary unit, then J is orthogonal. For $x \in \mathbb{C}, |x| \neq 0$, the vectors (x, Jx) form a normalized orthogonal basis. The map F is conformal if and only if (Fx, FJx) is again normalized orthogonal. On the other hand (Fx, JFx) *is* normalized orthogonal. Hence F is conformal, if and only if

$$ FJ = \pm JF, $$

where the sign depends on the orientation behaviour of F.

Note that this condition does not involve the scalar product, but only involves the complex structure J. A generalization of this fact to quaternions is fundamental for the theory presented here.

If $F : \mathbb{R}^2 = \mathbb{C} \to \mathbb{R}^4 = \mathbb{H}$ is $\mathbb{R}$-linear and injective, then $U = F(\mathbb{R}^2)$ is a real 2-dimensional subspace of $\mathbb{H}$, oriented by J. Let $N, R \in \mathbb{H}$ be its left and right normal vectors. Then $NU = U = UR$, and N induces an orthogonal endomorphism of U compatible with the Euclidean scalar product of $\mathbb{R}^4$. The map $F : \mathbb{R}^2 \to U$ is conformal if and only if $FJ = NF$. Hence $F : \mathbb{C} \to \mathbb{H}$ is conformal if and only if there exist $N, R \in \mathbb{H}, N^2 = -1 = R^2$, such that

$$*F := FJ = NF = -FR.$$

This leads to the following fundamental

Definition 2. *Let M be a Riemann surface, i.e. a 2-dimensional manifold endowed with a complex structure $J : TM \to TM, J^2 = -I$. A map $f : M \to \mathbb{H} = \mathbb{R}^4$ is called* conformal, *if there exist $N, R : M \to \mathbb{H}$ such that with $*df := df \circ J$,*

$$N^2 = -1 = R^2 \tag{2.4}$$

$$\boxed{*df = Ndf = -dfR.} \tag{2.5}$$

If f is an immersion then (2.4) follows from (2.5), and N and R are unique, called the left *and* right *normal vector of f.*

Remark 2. – Equation (2.5) is an analog of

$$*df = idf$$

for functions $f : \mathbb{C} \to \mathbb{C}$, i.e. of the Cauchy-Riemann equations. In this sense conformal maps into $\mathbb{H}$ are a generalization of complex holomorphic maps.

– If f is an immersion, then $df(T_pM) \subset \mathbb{H}$ is a 2-dimensional real subspace. Hence, according to Lemma 2, there exist N, R, inducing a complex structure J on $T_pM \cong df(T_pM)$. The definition requires that J coincides with the complex structure already given on T_pM.

– For an immersion f the existence of $N : M \to \mathbb{H}$ such that $*df = Ndf$ already implies that the immersion $f : M \to \mathbb{H}$ is conformal. Similarly for R.

– If $f : M \to \text{Im}\,\mathbb{H} = \mathbb{R}^3$ is an immersion then $N = R$ is "the classical" unit normal vector of f. But for general $f : M \to \mathbb{H}$, the vectors N and R are *not* orthogonal to $df(TM)$.

3 Projective Spaces

In complex function theory the Riemann sphere $\mathbb{C}P^1$ is more convenient as a target space for holomorphic functions than the complex plane. Similarly, the natural target space for conformal immersions is $\mathbb{H}P^1$, rather than $\mathbb{H}$. We therefore give a description of the quaternionic projective space.

3.1 Projective Spaces and Affine Coordinates

The quaternionic projective space $\mathbb{H}P^n$ is defined, similar to its real and complex cousins, as the set of quaternionic lines in $\mathbb{H}^{n+1}$. We have the (continuous) canonical projection

$$\pi : \mathbb{H}^{n+1} \backslash \{0\} \to \mathbb{H}P^n, x \mapsto \pi(x) = [x] = x\mathbb{H}.$$

The manifold structure of $\mathbb{H}P^n$ is defined as follows:
For any linear form $\beta \in (\mathbb{H}^{n+1})^*, \beta \neq 0$,

$$u : \pi(x) \mapsto x < \beta, x >^{-1}$$

is well-defined and maps the open set $\{\pi(x) \,|\, < \beta, x > \neq 0\}$ onto the affine hyperplane $\beta = 1$, which is isomorphic to $\mathbb{H}^n$. Coordinates of this type are called *affine coordinates* for $\mathbb{H}P^n$. They define a (real-analytic) atlas for $\mathbb{H}P^n$.

We shall often use this in the following setting: We choose a basis for $\mathbb{H}^{n+1}$ such that β is the last coordinate function. Then we get

$$\begin{bmatrix} x_1 \\ \vdots \\ x_n \\ x_{n+1} \end{bmatrix} \mapsto \begin{pmatrix} x_1 x_{n+1}^{-1} \\ \vdots \\ x_n x_{n+1}^{-1} \\ 1 \end{pmatrix} \text{ or } \begin{pmatrix} x_1 x_{n+1}^{-1} \\ \vdots \\ x_n x_{n+1}^{-1} \end{pmatrix}$$

The set

$$\{\pi(x) \,|\, < \beta, x >= 0\}$$

is called the *hyperplane at infinity*.

Example 2. In the special case $n = 1$, the hyperplane at infinity is a single point: $\mathbb{H}P^1$ is the one-point compactification of $\mathbb{R}^4$, hence "the" 4-sphere:

$$\boxed{\mathbb{H}P^1 = S^4.}$$

Note however, that the notion of *the antipodal map* is natural on the usual 4-sphere, but not on $\mathbb{H}P^1$ – unless we introduce additional structure, like a metric.

For our purposes it is important to have a good description of the tangent space $T_l \mathbb{H}P^n$ for $l \in \mathbb{H}P^n$. For that purpose, we consider the projection

$$\pi : \mathbb{H}^{n+1} \backslash \{0\} \to \mathbb{H}P^n$$

in affine coordinates: If $\beta \in (\mathbb{H}^{n+1})^*$ is as above, then

$$h = u \circ \pi : \mathbb{H}^{n+1} \backslash \{0\} \to \mathbb{H}^{n+1}, x \to x < \beta, x >^{-1}$$

satisfies

$$d_x h(v) = v < \beta, x >^{-1} -x < \beta, x >^{-1} < \beta, v >< \beta, x >^{-1} .$$

Therefore

$$\ker d_x h = x\mathbb{H},$$
$$d_{x\lambda} h(v\lambda) = d_x h(v)$$

for $\lambda \in \mathbb{H} \backslash \{0\}$, and the same holds for π:

$$\ker d_x \pi = x\mathbb{H}, \tag{3.1}$$
$$d_{x\lambda} \pi(v\lambda) = d_x \pi(v). \tag{3.2}$$

By (3.1), $d_x \pi$ induces an isomorphism

$$d_x \pi : \mathbb{H}^{n+1} / l \overset{\cong}{\to} T_l \mathbb{H}P^n, \quad l = \pi(x),$$

of real vector spaces, but it depends on the choice of $x \in l$. To eliminate this dependence, we note that by (3.2) the map

$$\text{Hom}(l, \mathbb{H}^{n+1} / l) \to T_l \mathbb{H}P^n, F \mapsto d_x \pi(F(x)),$$

with $x \in l \backslash \{0\}$ is a well-defined isomorphism:

$$\text{Hom}(l, \mathbb{H}^{n+1} / l) \cong T_l \mathbb{H}P^n. \tag{3.3}$$

In other words, we identify $d_x \pi(v)$ with the homomorphism from $l = \pi(x) = x\mathbb{H}$ to $\mathbb{H}^{n+1} / l$ that maps x to $\pi_l(v) := v \mod l$. For practical use, we rephrase this as follows:

Proposition 1. *Let $\tilde{f} : M \to \mathbb{H}^{n+1} \backslash \{0\}$ and $f = \pi\tilde{f} : M \to \mathbb{H}P^n$. Let $p \in M, l := f(p), v \in T_pM$. Then*

$$d_p f : T_p M \to T_{f(p)}\mathbb{H}P^n = \mathrm{Hom}(f(p), \mathbb{H}^{n+1}/f(p))$$

is given by

$$d_p f(v)(\tilde{f}(p)\lambda) = \pi_l(d_p\tilde{f}(v)\lambda).$$

We denote the differential in this interpretation by δf:

$$\boxed{\delta f(v)(\tilde{f}) = d\tilde{f}(v) \mod f.} \tag{3.4}$$

Proof. The tangent vector

$$d_p f(v) = d_{\tilde{f}(p)}\pi(d_p\tilde{f}(v)) \in T_{f(p)}\mathbb{H}P^n$$

is identified with the homomorphism $F : f(p) \to \mathbb{H}^n/f(p)$, that maps $\tilde{f}(p)$ into $d_p\tilde{f}(v) \mod f(p)$.

3.2 Metrics on $\mathbb{H}P^n$

Given a non-degenerate quaternionic hermitian inner product $< .,. >$ on $\mathbb{H}^{n+1}$, we define a (possibly degenerate Pseudo-) Riemannian metric on $\mathbb{H}P^n$ as follows: For $x \in \mathbb{H}^{n+1}$ with $< x, x > \neq 0$ and $v, w \in (x\mathbb{H})^\perp$ we define

$$< d_x\pi(v), d_x\pi(w) > = \frac{1}{< x, x >} \mathrm{Re} < v, w > .$$

This is well-defined since, for $0 \neq \lambda \in \mathbb{H}$, we have

$$< d_{x\lambda}\pi(v\lambda), d_{x\lambda}\pi(w\lambda) > = < d_x\pi(v), d_x\pi(w) > .$$

It extends to arbitrary v, w by

$$< d_x\pi(v), d_x\pi(w) > = \mathrm{Re} \frac{< v, w - x < x, w > < x, x >^{-1} >}{< x, x >}$$

$$= \mathrm{Re} \frac{< v, w > < x, x > - < v, x > < x, w >}{< x, x >^2}. \tag{3.5}$$

Example 3. For $< v, w > = \sum \bar{v}_k w_k$ we obtain the standard Riemannian metric on $\mathbb{H}P^n$. (In the complex case, this is the so-called Fubini-Study metric.) The corresponding conformal structure is in the background of all of the following considerations.

We take this standard Riemannian metric on $\mathbb{H}P^1 = S^4$ and ask which metric it induces on $\mathbb{R}^4$ via the affine parameter

$$h : \mathbb{H} \to \mathbb{H}P^1, x \mapsto \begin{bmatrix} x \\ 1 \end{bmatrix}.$$

Let $\tilde{h} : \mathbb{H} \to \mathbb{H}^2, x \mapsto (x, 1)$, and let "$\equiv$" denote equality mod $\begin{pmatrix} x \\ 1 \end{pmatrix} \mathbb{H}$.

Then

$$\delta_x h(v)(\begin{pmatrix} x \\ 1 \end{pmatrix})) \equiv d_x \tilde{h}(v) \equiv \begin{pmatrix} v \\ 0 \end{pmatrix} \equiv \begin{pmatrix} v \\ 0 \end{pmatrix} - \begin{pmatrix} x \\ 1 \end{pmatrix} \frac{\bar{x}v}{1 + x\bar{x}} \equiv \begin{pmatrix} v \\ -\bar{x}v \end{pmatrix} \frac{1}{1 + x\bar{x}}.$$

The latter vector is $< ., . >$-orthogonal to $(x, 1)$, and therefore the induced metric on $\mathbb{H}$ is given by

$$h^* < v, w >_x = \frac{1}{(1 + x\bar{x})^3} \operatorname{Re} < \begin{pmatrix} v \\ -\bar{x}v \end{pmatrix}, \begin{pmatrix} w \\ -\bar{x}w \end{pmatrix} >$$

$$= \frac{1}{(1 + x\bar{x})^2} \operatorname{Re}(\bar{v}w) = \frac{1}{(1 + x\bar{x})^2} < v, w >_{\mathbb{R}}.$$

But stereographic projection of S^4 induces the metric

$$\frac{4}{(1 + x\bar{x})^2} < v, w >_{\mathbb{R}}$$

on $\mathbb{R}^4$. Hence the standard metric on $\mathbb{H}P^1$ is of constant curvature 4.

Example 4. If we consider an *indefinite* hermitian metric on $\mathbb{H}^{n+1}$, then the above construction of a metric on $\mathbb{H}P^n$ fails for isotropic lines ($< l, l >= 0$), but these points are scarce. We consider the case $n = 1$, and the hermitian inner product

$$< v, w >= \bar{v}_1 w_2 + \bar{v}_2 w_1.$$

Isotropic lines are characterized in affine coordinates $h : x \mapsto \begin{pmatrix} x \\ 1 \end{pmatrix}$ by

$$0 =< \begin{pmatrix} x \\ 1 \end{pmatrix}, \begin{pmatrix} x \\ 1 \end{pmatrix} >= \bar{x} + x,$$

i.e. by $x \in \operatorname{Im} \mathbb{H} = \mathbb{R}^3$.

The point at infinity $\begin{pmatrix} 1 \\ 0 \end{pmatrix} \mathbb{H}$ is isotropic, too. Therefore, the set of isotropic points is a 3-sphere $S^3 \subset S^4$, and its complement consists of two open discs or – in affine coordinates – two open half-spaces.

As in the previous example, we find

$$h^* < v, w >_x = \frac{1}{(2 \operatorname{Re} x)^2} \operatorname{Re}(\bar{v}w) = \frac{1}{(2 \operatorname{Re} x)^2} < v, w >_{\mathbb{R}}$$

for the induced metric on the half-spaces $\operatorname{Re} \neq 0$ of $\mathbb{H}$. This is – up to a constant factor – the standard hyperbolic metric on these half-spaces.

3.3 Moebius Transformations on $\mathbb{H}P^1$

The group $Gl(2, \mathbb{H})$ acts on $\mathbb{H}P^1$ by $G(v\mathbb{H}) := Gv\mathbb{H}$. The kernel of this action, i.e. the set of all $G \in Gl(2, \mathbb{H})$ such that $Gv \in v\mathbb{H}$ for all v, is $\{\rho I \mid \rho \in \mathbb{R}\}$.

How is this action compatible with the metric induced by a positive definite hermitian metric of $\mathbb{H}^2$? Using (3.5) we find

$$|dG(d_x\pi(v\lambda))|^2 = \text{Re} \frac{<G(v\lambda), G(v\lambda)><Gx, Gx> - <G(v\lambda), Gx><Gx, G(v\lambda)>}{<Gx, Gx>^2}$$

$$= |\lambda|^2 \text{Re} \frac{<Gv, Gv><Gx, Gx> - <Gv, Gx><Gx, Gv>}{<Gx, Gx>^2}$$

$$= |\lambda|^2 |dG(d_x\pi(v))|^2$$

Taking $G = I$ we see that for $v \neq 0$ the map

$$\mathbb{H} \to T_{\pi(x)}\mathbb{H}P^1, \lambda \mapsto d_x\pi(v\lambda)$$

is length-preserving up to a constant factor, i.e. is a conformal isomorphism. But the same is obviously true for the metric induced by the pull-back under an arbitrary G, and therefore $GL(2, \mathbb{H})$ acts conformally on $\mathbb{H}P^1 = S^4$. We call these transformations the *Moebius transformations* on $\mathbb{H}P^1$. In affine coordinates they are given by

$$\begin{pmatrix} a & b \\ c & d \end{pmatrix} \begin{bmatrix} x \\ 1 \end{bmatrix} = \begin{bmatrix} ax + b \\ cx + d \end{bmatrix} = \begin{bmatrix} (ax + b)(cx + d)^{-1} \\ 1 \end{bmatrix}.$$

This emphasises the analogy with the complex case.

It is known that this is the full group of all orientation preserving conformal diffeomorphisms of S^4, see [7].

3.4 Two-Spheres in S^4

We consider the set

$$\mathcal{Z} = \{S \in \text{End}(\mathbb{H}^2) \mid S^2 = -I\}.$$

For $S \in \mathcal{Z}$ we define

$$S' := \{l \in \mathbb{H}P^1 \mid Sl = l\}.$$

We want to show

Proposition 2. *1. S' is a 2-sphere in $\mathbb{H}P^1$, i.e corresponds to a real 2-plane in $\mathbb{H} = \mathbb{R}^4$ under a suitable affine coordinate.*

  2. Each 2-sphere can be obtained in this way by an $S \in \mathcal{Z}$, unique up to sign.

Proof. We consider $\mathbb{H}^2$ as a (right) complex vector space with imaginary unit i. Then S is $\mathbb{C}$-linear and has a (complex) eigenvalue N. If $Sv = vN$, then

$$S(v\mathbb{H}) = vN\,\mathbb{H} = v\mathbb{H}.$$

Hence $S' \neq \emptyset$.

We choose a basis v, w of $\mathbb{H}^2$ such that $v\mathbb{H} \in S'$, i.e. $Sv = vN$ for some N, and $Sw = -vH - wR$. Then $S^2 = -I$ implies

$$N^2 = -1 = R^2, \quad NH = HR.$$

For the affine parametrization $h : \mathbb{H} \to \mathbb{H}P^1, x \mapsto [vx + w]$ we get:

$$
\begin{aligned}
[vx + w] \in S' &\iff \exists_\gamma \; S(vx + w) = (vx + w)\gamma \\
&\iff \exists_\gamma \; vNx - vH - wR = vx\gamma + w\gamma \\
&\iff \exists_\gamma \; \begin{cases} Nx - H = x\gamma \\ -R = \gamma \end{cases} \\
&\iff Nx + xR = H.
\end{aligned}
$$

This is a real-linear equation for x, with associated homogeneuos equation

$$Nx + xR = 0.$$

By Lemma 2 this is of real dimension 2, and any real 2-plane can be realized this way.

Obviously, S and $-S$ define the same 2-sphere. But S determines (N, R), thus fixing an orientation of the above real 2-plane and thereby of S'. Hence the lemma can be paraphrased as follows:

$\mathcal{Z}$ is the set of oriented 2-spheres in $S^4 = \mathbb{H}P^1$.

4 Vector Bundles

We shall need vector bundles over the quaternions, and therefore briefly introduce them.

4.1 Quaternionic Vector Bundles

A quaternionic vector bundle $\pi : V \to M$ of rank n over a smooth manifold M is a real vector bundle of rank $4n$ together with a smooth fibre-preserving action of $\mathbb{H}$ on V from the right such that the fibres become quaternionic vector spaces.

Example 5. The product bundle $\pi : M \times \mathbb{H}^n \to M$ with the projection on the first factor and the obvious vector space structure on each fibre $\{x\} \times \mathbb{H}^n$ is also called *the trivial bundle*.

Example 6. The points of the projective space $\mathbb{H}P^n$ are the 1-dimensional subspaces of $\mathbb{H}^{n+1}$. The *tautological bundle*

$$\pi_\Sigma : \Sigma \to \mathbb{H}P^n$$

is the line bundle with $\Sigma_l = l$. More precisely

$$\Sigma := \{(l, v) \in \mathbb{H}P^n \times \mathbb{H}^{n+1} \mid v \in l\},$$
$$\pi_\Sigma : \Sigma \to \mathbb{H}P^n, (l, v) \mapsto l.$$

The differentiable and vector space structure are the obvious ones.

Example 7. If $V \to \tilde{M}$ is a quaternionic vector bundle over $\tilde{M}$, and $f : M \to \tilde{M}$ is a map, then the "pull-back" $f^*V \to M$ is defined by

$$f^*V := \{(x, v) \mid v \in V_{f(x)}\} \subset M \times V$$

with the obvious projection and vector bundle structure. The fibre over $x \in M$ is just the fibre of V over $f(x)$.

We shall be concerned with maps $f : M \to \mathbb{H}P^n$ from a surface into the projective space. To f we associate the bundle $L := f^*\Sigma$, whose fibre over x is $f(x) \subset \mathbb{H}^{n+1} = \{x\} \times \mathbb{H}^{n+1}$. The bundle L is a line subbundle of the product bundle

$$H := M \times \mathbb{H}^{n+1}.$$

Conversely, every line subbundle L of H over M determines a map $f : M \to \mathbb{H}P^n$ by $f(x) := L_x$. We obtain an identification

Maps $f : M \to \mathbb{H}P^n$	$\leftrightarrow$	Line subbundles $L \subset H = M \times \mathbb{H}^{n+1}$

All natural constructions for vector spaces extend, fibre-wise, to operations in the category of vector bundles. For example, a subbundle L of a vector bundle H induces a quotient bundle H/L with fibres H_x/L_x. Given two quaternionic vector bundles V_1, V_2 the *real* vector bundle $\mathrm{Hom}(V_1, V_2)$ has the fibres $\mathrm{Hom}(V_{1x}, V_{2x})$. A section $\Phi \in \Gamma(\mathrm{Hom}(V_1, V_2))$ is called a vector bundle homomorphism. It is a smooth map $\Phi : V_1 \to V_2$ such that for all x the restriction $\Phi|_{V_{1x}}$ maps V_{1x} homomorphically into V_{2x}. There is an obvious notion of *isomorphism* for vector bundles.

Example 8. Over $\mathbb{H}P^n$ we have the product bundle $H = \mathbb{H}P^n \times \mathbb{H}^{n+1}$ and, inside it, the tautological subbundle Σ. Then

$$T\mathbb{H}P^n \cong \mathrm{Hom}(\Sigma, H/\Sigma),$$

see (3.3).

Example 9 (and Definition). Let L be a line subbundle of $H = M \times \mathbb{H}^{n+1}$. Let $\pi_L : H \to H/L \in \Gamma(\mathrm{Hom}(H, H/L))$ be the projection. A section $\psi \in \Gamma(L) \subset \Gamma(H)$ is a particular map $\psi : M \to \mathbb{H}^{n+1}$. If $X \in T_pM$, then $d\psi(X) \in H_p = \mathbb{H}^{n+1}$, and

$$\pi_L(d\psi(X)) \in (H/L)_p = \mathbb{H}^{n+1}/L_p.$$

Let $\lambda : M \to \mathbb{H}$. Then

$$\pi_L(d(\psi\lambda)(X)) = \pi_L(d\psi(X)\lambda + \psi d\lambda(X)) = \pi_L(d\psi(X))\lambda.$$

We see that

$$\psi \mapsto \pi_L(d\psi(X)) =: \delta(X)(\psi)$$

is tensorial in ψ, i.e. we obtain

$$\delta(X) = \delta_L(X) \in \mathrm{Hom}(L_p, (H/L)_p).$$

Of course this is $\mathbb{R}$-linear in X as well, so δ should be viewed as a 1-form on M with values in $\mathrm{Hom}(L, H/L)$:

$$\boxed{\delta \in \Omega^1(\mathrm{Hom}(L, H/L)).} \tag{4.1}$$

Let us repeat: Given $p \in M, X \in T_p M$, and $\psi_0 \in L_p$, there is a section $\psi \in \Gamma(L)$ such that $\psi(p) = \psi_0$. Then

$$\boxed{\delta_p(X)\psi_0 = \pi_L(d_p\psi(X)) = d_p\psi(X) \quad \mod L_p.}$$

Note the similarity to the second fundamental form

$$\alpha(X, Y) = (dY(X))^\perp.$$

of a submanifold M in Euclidan space. In the case at hand, L corresponds to TM and $\mathbb{H}^{n+1}/L$ corresponds to the normal bundle. This is *the* general method to measure the change of a subbundle L in a (covariantly connected) vector bundle H.

We can view L as a map $f : M \to \mathbb{H}P^n$. Even if this is an immersion, δ clearly has nothing to do with the second fundamental form of f. Instead, comparison with Proposition 1 shows that

$$\delta : TM \to \mathrm{Hom}(L, H/L)$$

corresponds to the derivative of f, and we shall therefore call it the *derivative of L* .

Example 10. The dual $V^* := \{\omega : V \to \mathbb{H} | \omega \ \mathbb{H}\text{-linear}\}$ of a quaternionic vector space V is, in a natural way, a *left* $\mathbb{H}$-vector space. But since we choose quaternionic vector spaces to be *right* vector spaces, we use the opposite structure: For $\omega \in V^*$ and $\lambda \in \mathbb{H}$ we define

$$\omega\lambda := \bar{\lambda}\omega.$$

This extends to quaternionic vector bundles. E.g., if L is a line bundle, i.e. of rank 1, then L^* is another quaterionic line bundle, usually denoted by L^{-1}.

A quaternionic vector bundle is called trivial if it is isomorphic with the product bundle $M \times \mathbb{H}^n$, i.e. if there exist global sections $\phi_1, \ldots, \phi_n : M \to V$ that form a basis of the fibre everywhere. Note that for a quaternionic *line* bundle over a surface the total space V has real dimension $2 + 4 = 6$, and hence any section $\phi : M \to V$ has codimension 4. It follows from transversality theory that any section can be slightly deformed so that it will not hit the 0-section. Therefore there exists a global nowhere vanishing section: Any quaternionic line bundle over a Riemann surface is (topologically) trivial.

4.2 Complex Quaternionic Bundles

A *complex quaternionic vector bundle* is a pair (V, J) consisting of a quaternionic vector bundle V and a section $J \in \Gamma(\operatorname{End}(V))$ with

$$J^2 = -I,$$

see section 2.1.

Example 11. Given $f : M \to \mathbb{H}, *df = Ndf$, the quaternionic line bundle $L = M \times \mathbb{H}$ has a complex structure given by

$$Jv := Nv.$$

Example 12. For a given $S \in \operatorname{End}(\mathbb{H}^2)$ with $S^2 = -I$, we identified

$$S' = \{l \mid Sl = l\} \subset \mathbb{H}P^1$$

as a 2-sphere in $\mathbb{H}P^1$, see section 3.4. We now compute δ, or rather the image of δ, for the corresponding line bundle L. In other words, we compute the tangent space of $S' \subset \mathbb{H}P^1$.

Note that, because of $SL \subset L$, S induces a complex structure on L, and it also induces one (again denoted by S) on H/L such that $\pi_L S = S\pi_L$. Now for $\psi \in \Gamma(L)$, we have

$$\delta S\psi = \pi_L d(S\psi) = \pi_L S d\psi = S\pi_L d\psi = S\delta\psi.$$

This shows

$$TS' = \operatorname{image}\delta \subset \operatorname{Hom}_+(L, H/L).$$

But the real vector bundle $\operatorname{Hom}_+(L, H/L)$ has rank 2, see Example 1, and since S' is an embedded surface, the inclusion is an equality:

$$T_l S' = \operatorname{Hom}_+(L_l, (H/L)_l) \subset \operatorname{Hom}(L_l, (H/L)_l) = T_l \mathbb{H}P^1.$$

For our next example we generalize Lemma 2.

Lemma 3. *Let V, W be 1-dimensional quaternionic vector spaces, and*

$$U \subset \operatorname{Hom}(V, W)$$

be a 2-dimensional real vector subspace. Then there exists a pair of complex structures $J \in \operatorname{End}(V), \tilde{J} \in \operatorname{End}(W)$, unique up to sign, such that

$$\tilde{J}U = U = UJ,$$
$$U = \{F \in \operatorname{Hom}(V, W) \mid \tilde{J}FJ = -F\}$$

If U is oriented, then there is only one such pair such that J is compatible with the orientation.

Note: Here we choose the sign of J in such a way that it corresponds to $-R$ rather than R.

Proof. Choose non-zero basis vectors $v \in V, w \in W$. Then elements in $\mathrm{Hom}(V, W)$ and endomorphisms of V or of W are represented by quaternionic 1×1-matrices, and therefore the assertion reduces to that of Lemma 2.

The following is now evident:

Proposition 3. *Let $L \subset H = M \times \mathbb{H}^2$ be an immersed oriented surface in $\mathbb{H}P^1$ with derivative $\delta \in \Omega^1(\mathrm{Hom}(L, H/L))$. Then there exist unique complex structures on L and H/L, denoted by $J, \tilde{J}$, such that for all $x \in M$*

$$\tilde{J}\delta(T_x M) = \delta(T_x M) = \delta(T_x M)J,$$
$$\tilde{J}\delta = \delta J,$$

and J is compatible with the orientation induced by $\delta : T_x M \to \delta(T_x M)$.

Definition 3. *A line subbundle $L \subset H = M \times \mathbb{H}^{n+1}$ over a Riemann surface M is called* conformal *or a holomorphic curve in $\mathbb{H}P^n$, if there exists a complex structure J on L such that*

$$*\delta = \delta J.$$

From the proposition we see: If L is an *immersed* holomorphic curve in $\mathbb{H}P^1$, i.e. if δ is in addition injective, such that $\delta(TM) \subset \mathrm{Hom}(L, H/L)$ is a real subbundle of rank 2, then there is also a complex structure $\tilde{J} \in \Gamma(\mathrm{End}(H/L))$ such that

$$*\delta = \delta J = \tilde{J}\delta. \tag{4.2}$$

A Riemann surface immersed into $\mathbb{H}P^1$ is a holomorphic curve if and only if the complex structures given by the proposition are compatible with the complex structure given on M in the sense of (4.2).

Example 13. Let $f : M \to \mathbb{H}$ be a conformally immersed Riemann surface with right normal vector R, and let L be the line bundle corresponding to

$$\begin{bmatrix} f \\ 1 \end{bmatrix} : M \to \mathbb{H}P^1.$$

Then $\begin{pmatrix} f \\ 1 \end{pmatrix} \in \Gamma(L)$, and

$$\delta(\begin{pmatrix} f \\ 1 \end{pmatrix} R) = \pi_L d(\begin{pmatrix} f \\ 1 \end{pmatrix} R) = \pi_L(\begin{pmatrix} df \\ 0 \end{pmatrix} R + \begin{pmatrix} f \\ 1 \end{pmatrix} dR)$$
$$= \pi_L \begin{pmatrix} dfR \\ 0 \end{pmatrix} = -\pi_L \begin{pmatrix} *df \\ 0 \end{pmatrix} = -*\delta \begin{pmatrix} f \\ 1 \end{pmatrix}.$$

If we define $J \in \mathrm{End}(L)$ by $J \begin{pmatrix} f \\ 1 \end{pmatrix} = -\begin{pmatrix} f \\ 1 \end{pmatrix} R$ then

$$\delta J = *\delta,$$

hence (L, J) is a holomorphic curve. Conversely, if (L, J) is a holomorphic curve, then $J\begin{pmatrix} f \\ 1 \end{pmatrix} = -\begin{pmatrix} f \\ 1 \end{pmatrix} R$ for some $R : M \to \mathbb{H}$, and f is conformal with right normal vector R.

4.3 Holomorphic Quaternionic Bundles

Let (V, J) be a complex quaternionic vector bundle over the Riemann surface M. We decompose

$$\operatorname{Hom}_{\mathbb{R}}(TM, V) = KV \oplus \bar{K}V,$$

where

$$KV := \{\omega : TM \to V \mid *\omega = J\omega\},$$
$$\bar{K}V := \{\omega : TM \to V \mid *\omega = -J\omega\}.$$

Definition 4. *A holomorphic structure on (V, J) is a quaternionic linear map*

$$D : \Gamma(V) \to \Gamma(\bar{K}V)$$

such that for all $\psi \in \Gamma(V)$ and $\lambda : M \to \mathbb{H}$

$$D(\psi\lambda) = (D\psi)\lambda + \frac{1}{2}(\psi d\lambda + J\psi * d\lambda). \tag{4.3}$$

A section $\psi \in \Gamma(V)$ is called holomorphic *if $D\psi = 0$, and we put*

$$H^0(V) = \ker D \subset \Gamma(V).$$

Remark 3.   1. For a better understanding of this, note that for complex-valued λ the anti-$\mathbb{C}$-linear part (the $\bar{K}$-part) of $d\lambda$ is given by $\bar{\partial}\lambda = \frac{1}{2}(d\lambda + i * d\lambda)$. In fact,

$$(d\lambda + i * d\lambda)(JX) = *d\lambda(X) - i\, d\lambda(X) = -i(d\lambda + i * d\lambda)(X).$$

A holomorphic structure is a generalized $\bar{\partial}$-operator. Equation (4.3) is the only natural way to make sense of a product rule of the form "$D(\psi\lambda) = (D\psi)\lambda + \psi\bar{\partial}\lambda$".

  2. If L is a holomorphic curve in $\mathbb{H}P^1$, does this mean L carries a natural holomorphic structure? This is not yet clear, but we shall come back to this question. See also Theorem 1 below.

Example 14. Any given $J \in \mathrm{End}(\mathbb{H}^n), J^2 = -1$, turns $H = M \times \mathbb{H}^n$ into a complex quaternionic vector bundle. Then $\Gamma(H) = \{\psi : M \to \mathbb{H}^n\}$, and

$$D\psi := \frac{1}{2}(d\psi + J * d\psi)$$

is a holomorphic structure.

Example 15. If L is a complex quaternionic line bundle and $\phi \in \Gamma(L)$ has no zeros, then there exists exactly one holomorphic structure D on (L, J) such that ϕ becomes holomorphic. In fact, any $\psi \in \Gamma(L)$ can be written as $\psi = \phi\mu$ with $\mu : M \to \mathbb{H}$, and our only chance is

$$D\psi := \frac{1}{2}(\phi d\mu + J\phi * d\mu). \tag{4.4}$$

This, indeed, satisfies the definition of a holomorphic structure.

Example 16. If $f : M \to \mathbb{H}$ is a conformal surface with left normal vector N, then N is a complex structure for $L = M \times \mathbb{H}$, and there exists a unique D such that $D1 = 0$. A section $\psi = 1\mu$ is holomorphic if and only if $d\mu + N * d\mu = 0$, i.e.

$$*d\mu = Nd\mu.$$

The holomorphic sections are therefore the conformal maps with the same left normal N as f. In this case $\dim H^0(L) \geq 2$, since 1 and f are independent in $H^0(L)$.

Theorem 1. *If $L \subset H = M \times \mathbb{H}^{n+1}$ is a holomorphic curve with complex structure J, then the dual bundle L^{-1} inherits a complex structure defined by $J\omega := \omega J$. The pair (L^{-1}, J) has a canonical holomorphic structure D characterized by the following fact: Any quaternionic linear form $\omega : \mathbb{H}^{n+1} \to \mathbb{H}$ induces a section $\omega_L \in \Gamma(L^{-1})$ by restriction to the fibres of L. Then for all ω*

$$D\omega_L = 0.$$

Proof. The vector bundle $L^\perp$ with fibre $L_x^\perp = \{\omega \in (\mathbb{H}^{n+1})^* \mid \omega|_{L_x} = 0\}$ has a total space of real dimension $4n + 2$. Therefore there exists ω such that ω_L has no zero. Example 15 yields a unique holomorphic structure D such that $D\omega_L = 0$. Now any $\alpha \in \Gamma(L^{-1})$ is of the form $\alpha = \omega_L \lambda$ for some $\lambda : M \to \mathbb{H}$. Then, by (4.4), for any section $\psi \in \Gamma(L)$ we have

$$< D\alpha, \psi > = \frac{1}{2} < \omega_L d\lambda + J\omega_L * d\lambda, \psi >$$

$$= \frac{1}{2}(< \omega d\lambda, \psi > + < \omega * d\lambda, J\psi >)$$

$$= \frac{1}{2}(d < \omega\lambda, \psi > + *d < \omega\lambda, J\psi >) - \frac{1}{2} < \omega\lambda, d\psi + *d(J\psi) >$$

$$= \frac{1}{2}(d < \alpha, \psi > + *d < \alpha, J\psi >) - \frac{1}{2} < \omega\lambda, d\psi + *d(J\psi) > .$$

Note that $*\delta = \delta J$ implies $d\psi + *d(J\psi) \in \Gamma(L)$, and this allows us to replace $\omega\lambda$ by α in the last term as well:

$$< D\alpha, \psi > = \frac{1}{2}(d < \alpha, \psi > + * d < \alpha, J\psi >) - \frac{1}{2} < \alpha, d\psi + *dJ\psi > .$$

This contains no reference to ω, hence D is independent of the choice of ω such that ω_L has no zero. But the last equality shows $D\alpha = 0$ for *any* $\alpha = \omega_L$ with $\omega \in (\mathbb{H}^{n+1})^*$.

Remark 4. As we shall see in the next section, a holomorphic curve L in $\mathbb{H}P^1$ carries a natural holomorphic structure. In higher dimensional projective spaces this is no longer the case. Therefore L^{-1} rather than L plays a prominent role in higher codimension.

5 The Mean Curvature Sphere

5.1 S-Theory

Let M be a Riemann surface. Let

$$H := M \times \mathbb{H}^2$$

denote the product bundle over M, and let $S : M \to \text{End}(\mathbb{H}^2) \in \Gamma(\text{End}(H))$ with $S^2 = -I$ be a complex structure on H. We split the differential according to type:

$$d\psi = d'\psi + d''\psi,$$

where d' and d'' denote the $\mathbb{C}$-linear and anti-linear components, respectively:

$$*d' = Sd', \quad *d'' = -Sd''.$$

Explicitly,

$$d'\psi = \frac{1}{2}(d\psi - S * d\psi), \quad d''\psi = \frac{1}{2}(d\psi + S * d\psi).$$

So d'' is a holomorphic structure on (H, S), while d' is an anti-holomorphic structure, i.e. a holomorphic structure of $(H, -S)$.

In general $d(S\psi) \neq Sd\psi$, and we decompose further:

$$d' = \partial + A, \quad d'' = \bar{\partial} + Q,$$

where

$$\partial(S\psi) = S\partial\psi, \quad \bar{\partial}(S\psi) = S\bar{\partial}\psi,$$
$$AS = -SA, \quad QS = -SQ.$$

For example, we explicitly have

$$\bar{\partial}\psi = \frac{1}{2}(d''\psi - Sd''(S\psi)).$$

Then $\bar{\partial}$ defines a holomorphic structure and ∂ an anti-holomorphic structure on H, while A and Q are tensorial:

$$A \in \Gamma(K \operatorname{End}_-(H)), \quad Q \in \Gamma(\bar{K} \operatorname{End}_-(H)). \tag{5.1}$$

For $\psi : M \to \mathbb{H}^2 \in \Gamma(H)$ we have, by definition of dS,

$$
\begin{aligned}
(dS)\psi &= d(S\psi) - S d\psi \\
&= (\partial + A)S\psi + (\bar{\partial} + Q)S\psi - S(\partial + A)\psi - S(\bar{\partial} + Q)\psi \\
&= AS\psi + QS\psi - SA\psi - SQ\psi \\
&= -2S(Q + A)\psi \\
&= 2(*Q - *A)\psi.
\end{aligned}
$$

Hence

$$dS = 2(*Q - *A), \quad *dS = 2(A - Q). \tag{5.2}$$

Then

$$S dS = 2(Q + A),$$

whence conversely

$$Q = \frac{1}{4}(S dS - *dS), \quad A = \frac{1}{4}(S dS + *dS). \tag{5.3}$$

Remark 5. Since A and Q are of different type, $dS = 0$ if and only if $A = 0$ and $Q = 0$. If $dS = 0$, then the $\pm i$-eigenspaces of the complex endomorphism S decompose $H = (M \times \mathbb{C}) \oplus (M \times \mathbb{C})$. Therefore A and Q measure the deviation from the "complex case".

5.2 The Mean Curvature Sphere

We now consider an immersed holomorphic curve $L \subset H$ in $\mathbb{H}P^1$ with derivative $\delta = \delta_L \in \Omega^1(\operatorname{Hom}(L, H/L))$. Then there exist complex structures J on L and $\tilde{J}$ on H/L such that

$$*\delta = \delta J = \tilde{J}\delta.$$

We want to extend J and $\tilde{J}$ to a complex structure of H, i.e. find an

$$S \in \Gamma(\operatorname{End}(H))$$

such that

$$SL = L, \quad S|_L = J, \quad \pi S = \tilde{J}\pi.$$

Note that this implies

$$\pi dS(\psi) = \pi(d(S\psi) - Sd\psi) = \delta J\psi - \tilde{J}\delta\psi = 0,$$

and therefore

$$dSL \subset L. \tag{5.4}$$

The existence of S is clear: Write $H = L \oplus L'$ for some complementary bundle L'. Identify L' with H/L using π, and define $S|_L := J, S|_{L'} := \tilde{J}$. Since L' is not unique, S is not unique. It is easy to see that $\tilde{S} = S + R$ is another such extension if and only if $R : M \to \mathrm{End}(\mathbb{H}^2)$ satisfies

$$RH \subset L \subset \ker R,$$

whence $R^2 = 0$, and

$$RS + SR = 0.$$

Note that R can be interpreted as an element of $\mathrm{Hom}(H/L, L)$. Then $R\pi = R$. We compute $\tilde{Q}$:

$$\begin{aligned}
\tilde{Q} &= \frac{1}{4}((S + R)d(S + R) - *d(S + R)) \\
&= \frac{1}{4}(SdS - *dS) + \frac{1}{4}(SdR + RdS + RdR - *dR) \\
&= Q + \frac{1}{4}(SdR + RdS + RdR - *dR).
\end{aligned}$$

If $\psi \in \Gamma(L)$, then

$$\begin{aligned}
0 &= d(R\psi) = dR\psi + Rd\psi, \\
RdR\psi &= -R^2 d\psi = 0
\end{aligned}$$

and, by (5.4),

$$R \underbrace{dS\psi}_{\in \Gamma(L)} = 0$$

We can therefore continue

$$\begin{aligned}
\tilde{Q}\psi &= Q\psi + \frac{1}{4}(SdR\psi - *dR\psi) = Q\psi + \frac{1}{4}(-SRd\psi + *Rd\psi) \\
&= Q\psi + \frac{1}{4}(-SR\delta\psi + R * \delta\psi) = Q\psi + \frac{1}{4}(-SR\delta\psi + \underbrace{R\tilde{J}}_{=RS=-SR} \delta\psi).
\end{aligned}$$

Hence, for $\psi \in \Gamma(L)$,

$$\tilde{Q}\psi = Q\psi - \frac{1}{2}SR\delta\psi. \tag{5.5}$$

Now we start with any extension S of $(J, \tilde{J})$ and, in view of (5.5), define

$$R = -2SQ(X)\delta(X)^{-1}\pi : H \to H \qquad (5.6)$$

for some $X \neq 0$. First note that this definition is independent of the choice of $X \neq 0$. In fact, $X \mapsto R$ is positive-homegeneous of degree 0, and with $c = \cos\theta, s = \sin\theta$

$$Q(cX + sJX)(\delta(cX + sJX))^{-1}) = Q(X)(cI + sS)(\delta(X)(cI + sS))^{-1}$$
$$= Q(X)\delta(X)^{-1}.$$

Next

$$RS = -2SQ(X)\delta_X^{-1}\pi S = -2SQ(X)\delta_X^{-1}\tilde{J}\pi$$
$$= -2SQ(X)S\delta_X^{-1}\pi = 2S^2 Q(X)\delta_X^{-1}\pi$$
$$= -SR.$$

By definition (5.6)

$$L \subset \ker R,$$

and from (5.3) and (5.4) we get

$$L \supset \frac{1}{4}(SdS - *dS)L = QL,$$

whence

$$RH \subset L.$$

We have now shown that $\tilde{S} = S + R$ is another extension. Finally, using (5.5), we find for $\psi \in \Gamma(L)$

$$\tilde{Q}\psi = Q\psi - \frac{1}{2}SRd\psi = Q\psi - \frac{1}{2}S(-2SQ\delta^{-1}\pi)d\psi$$
$$= Q\psi - Q\delta^{-1}\pi d\psi = 0.$$

This shows

Theorem 2. *Let $L \subset H = M \times \mathbb{H}^2$ be a holomorphic curve immersed into $\mathbb{HP}^1$. Then there exists a unique complex structure S on H such that*

$$SL = L, \quad dSL \subset L, \qquad (5.7)$$
$$*\delta = \delta \circ S = S \circ \delta, \qquad (5.8)$$
$$Q|_L = 0. \qquad (5.9)$$

S is a family of 2-spheres, a *sphere congruence* in classical terms. Because $S_p L_p = L_p$ the sphere S_p goes through $L_p \in \mathbb{HP}^1$, while $dSL \subset L$ (or, equivalently, $\delta S = S\delta$) implies it is tangent to L in p, see examples 9 and 12. In an affine coordinate system $\begin{bmatrix} f \\ 1 \end{bmatrix} = L$ the sphere S_p has the same mean curvature vector as $f : M \to \mathbb{R}^4 = \mathbb{H}$ at p, see Remark 9. This motivates the

Definition 5. *S is called the* mean curvature sphere (congruence) *of L. The differential forms $A, Q \in \Omega^1(\mathrm{End}(H))$ are called* the Hopf fields *of L.*

Remark 6. Equations (5.7), (5.8) imply $d\psi + S * d\psi \in \Gamma(L)$ for $\psi \in \Gamma(L)$, whence $d'' = \bar{\partial} + Q = \frac{1}{2}(d + S * d)$ leaves L invariant. Hence an immersed holomorphic curve in $\mathbb{HP}^1$ is a holomorphic subbundle of (H, S, d'') and, in particular, is a holomorphic quaternionic vector bundle itself.

Example 17. Let $S \in \mathrm{End}(\mathbb{H}^2), S^2 = -I$. Then

$$S' = \{l \in \mathbb{HP}^1 \mid Sl = l\} \subset \mathbb{HP}^1$$

is a 2-sphere in $\mathbb{HP}^1$. Let L denote the corresponding line bundle and endow S' with the complex structure inherited from the immersion. Then the mean curvature sphere congruence of L is simply the constant map $S' \to \mathcal{Z}$ of value S: We have $SL = L$ by definition, and the constancy implies $dSL = \{0\} \subset L$ and $Q = \frac{1}{4}(SdS - *dS) = 0$.

5.3 Hopf Fields

In the following we shall frequently encounter differential forms. Note that the usual definition of the wedge product of 1-forms

$$\omega \wedge \theta(X, Y) = \omega(X)\theta(Y) - \omega(Y)\theta(X)$$

can be generalized verbatim to forms $\omega_i \in \Omega^1(V_i)$ with values in vector spaces or bundles V_i, provided there is a product $V_1 \times V_2 \to V$. Examples are the composition $\mathrm{End}(V) \times \mathrm{End}(V) \to \mathrm{End}(V)$ or the pairing between the dual V^* and V.

On a Riemann surface M, any 2-form $\sigma \in \Omega^2$ is completely determined by the quadratic form $\sigma(X, JX) =: \sigma(X)$, and we shall, for simplicity, often use the latter. As an example,

$$\omega \wedge \theta(X, JX) = \omega(X)\theta(JX) - \omega(JX)\theta(X)$$

will be written as

$$\omega \wedge \theta = \omega * \theta - *\omega \, \theta. \qquad (5.10)$$

We now collect some information about the Hopf fields and the mean curvature sphere congruence $S : M \to \mathcal{Z}$.

Lemma 4.

$$d(A + Q) = 2(Q \wedge Q + A \wedge A).$$

Proof. Recall from (5.2)

$$SdS = 2(A + Q).$$

Therefore, using $AS = -SA, QS = -SQ$,

$$\begin{aligned} d(A + Q) &= \frac{1}{2}d(SdS) = \frac{1}{2}(dS \wedge dS) \\ &= 2S(A + Q) \wedge S(A + Q) \\ &= 2(A \wedge A + A \wedge Q + Q \wedge A + Q \wedge Q). \end{aligned}$$

But $A \wedge Q = 0$ by the following *type argument*: Using that

A is "right $\bar{K}$", and Q "left $\bar{K}$"

we have

$$A \wedge Q = A * Q - *AQ = A(-SQ) - (-AS)Q = 0. \tag{5.11}$$

Similarly $Q \wedge A = 0$, because A is left K and Q is right K.

Lemma 5. *Let $L \subset H$ be an immersed surface and S a complex structure on H stabilizing L such that $dSL \subset L$. Then $Q_{|L} = 0$ is equivalent to $AH \subset L$.*

Notice that the kernels and images of the 1-forms A and Q are well-defined: if $Q_X\psi = 0$ for some $X \in TM$ then also $Q_{JX}\psi = -SQ_X\psi = 0$, and thus $Q_Z\psi = 0$ for any $Z \in TM$. In other words, the kernels of Q and A are independent of $X \in TM$. The same remark holds for the respective images.

Proof. We first need a formula for the derivative of 1-forms $\omega \in \Omega^1(\text{End}(H))$ which stabilize L, i.e., $\omega L \subset L$. If $\pi = \pi_L$, then for $\psi \in \Gamma(L)$

$$\begin{aligned} \pi(d\omega(X,Y)\psi) &= \pi(d(\omega\psi)(X,Y) + \omega \wedge d\psi(X,Y)) \\ &= \pi(X \cdot (\omega(Y)\psi) - Y \cdot (\omega(X)\psi) - \underbrace{\omega([X,Y])\psi}_{\in \Gamma(L)} \\ &\quad + \omega(X)d\psi(Y) - \omega(Y)d\psi(X)) \\ &= \delta(X)\omega(Y)\psi - \delta(Y)\omega(X)\psi + \pi\omega(X)d\psi(Y) - \pi\omega(Y)d\psi(X) \\ &= \delta(X)\omega(Y)\psi - \delta(Y)\omega(X)\psi + \pi\omega(X)\delta\psi(Y) - \pi\omega(Y)\delta\psi(X) \\ &= (\delta \wedge \omega + \pi\omega \wedge \delta)(X,Y)\psi, \end{aligned}$$

where we wedge over composition. Note that the composition $\pi\omega\delta$ makes sense, because $\omega(L) \subset L$, and L is annihilated by π. We apply this to A and Q. Since $AL \subset L, QL \subset L$ we have <u>on L</u>, by lemma 4,

$$0 = \frac{1}{2}\pi(Q \wedge Q + A \wedge A) = \pi(dA + dQ)$$
$$= \delta \wedge A + \pi A \wedge \delta + \delta \wedge Q + \pi Q \wedge \delta.$$

By a type argument similar to (5.11), we get $\delta \wedge A = 0 = \pi Q \wedge \delta$. Further,

$$\pi A \wedge \delta = \pi A * \delta - \pi * A\delta$$
$$= -2S\pi A\delta,$$

and similarly for the remaining term. We obtain $-\pi S A\delta = S\delta Q|_L$ or

$$-\pi A\delta = \delta Q|_L.$$

Since $AL \subset L$ and $\delta(X) : L \to H/L$ for $X \neq 0$ is an isomorphism, we get $\pi A = 0 \iff Q|_L = 0$.

5.4 The Conformal Gauss Map

Definition 6. *For a quaternionic vector space or bundle V of rank n and $A \in \mathrm{End}(V)$ we define*

$$< A >:= \frac{1}{4n}\,\mathrm{trace}_{\mathbb{R}}\,A,$$

where the trace is taken of the real endomorphism A. In particular $< I >= 1$. We obtain an indefinite scalar product $< A, B >:=< AB >$.

Example 18. For $A = (a)$ with $a = a_0 + ia_1 + ja_2 + ka_3 \in \mathbb{H}$ we have

$$< A >= \frac{1}{4}4a_0 = a_0,$$

and

$$< AA >= \mathrm{Re}\,a^2 = a_0^2 - a_1^2 - a_2^2 - a_3^2.$$

Proposition 4. *The mean curvature sphere S of an immersed Riemann surface L satisfies*

$$< dS, dS >=< *dS, *dS >, \quad < dS, *dS >= 0,$$

i.e. $S : M \to \mathcal{Z}$ is conformal.

Because of this proposition, S is also called *the conformal Gauss map*, see Bryant [1].

Proof. We have $QA = 0$, and therefore

$$< Q, A >=< A, Q >= 0. \tag{5.12}$$

Then, from (5.2),

$$< dS, dS >=4 < -S(Q + A), -S(Q + A) >= 4 < Q + A, Q + A >$$
$$=4 < Q - A, Q - A >=< *dS, *dS > .$$

Similarly,

$$< dS, *dS >=4 < -S(Q + A), A - Q >$$
$$=4(< SQQ > - < S\underbrace{QA}_{=0} > + < SAQ > - < SAA >).$$

But, by a property of the real trace,

$$< SAQ > =< QSA >=< -SQA >= 0,$$
$$< SQQ > =< QSQ >=< -SQQ >= 0,$$
$$< SAA > =< ASA >=< -SAA >= 0.$$

6 Willmore Surfaces

Throughout this section M denotes a *compact* surface.

6.1 The Energy Functional

The set

$$\mathcal{Z} = \{S \in \text{End}(\mathbb{H}^2) \mid S^2 = -I\}$$

of oriented 2-spheres in $\mathbb{H}P^1$ is a submanifold of $\text{End}(\mathbb{H}^2)$ with

$$T_S \mathcal{Z} = \{X \in \text{End}(\mathbb{H}^2) \mid XS = -SX\},$$
$$\perp_S \mathcal{Z} = \{Y \in \text{End}(\mathbb{H}^2) \mid YS = SY\}.$$

Here we use the (indefinite) inner product

$$< A, B >:=< AB >= \frac{1}{8} \text{trace}_{\mathbb{R}}(AB)$$

defined in Section 5.3.

Definition 7. *The energy functional of a map* $S : M \to \mathcal{Z}$ *of a Riemann surface* M *is defined by*

$$E(S) := \int_M < dS \wedge *dS > .$$

Critical points S *of this functional with respect to variations of* S *are called harmonic maps from* M *to* $\mathcal{Z}$.

Proposition 5. S *is harmonic if and only if the* $\mathcal{Z}$-*tangential component of* $d * dS$ *vanishes:*

$$(d * dS)^T = 0. \tag{6.1}$$

This condition is equivalent to any of the following:

$$d(S * dS) = 0, \tag{6.2}$$
$$d * A = 0, \tag{6.3}$$
$$d * Q = 0. \tag{6.4}$$

In fact,

$$d(S * dS) = 4d * Q = 4d * A = S(d * dS)^T = (Sd * dS)^T. \tag{6.5}$$

Proof. Let S_t be a variation of S in $\mathcal{Z}$ with variational vector field $\dot{S} =: Y$. Then $SY = -YS$ and

$$\frac{d}{dt} E(S) = \frac{d}{dt} \int_M < dS \wedge *dS > = \int_M < dY \wedge *dS > + < dS \wedge *dY > .$$

Using the wedge formula (5.10) and $\text{trace}_{\mathbb{R}}(AB) = \text{trace}_{\mathbb{R}}(BA)$, we get

$$< dS \wedge *dY > = < dS(-dY) - *dS * dY > = < dY \wedge *dS > .$$

Thus

$$\frac{d}{dt} E(S) = 2 \int_M < dY \wedge *dS > = -2 \int_M < Yd * dS > = -2 \int_M < Y, d * dS > .$$

Therefore S is harmonic if and only if $d * dS$ is normal.

For the other equivalences, first note

$$0 = d * d(S^2) = d(*dSS + S * dS)$$
$$= (d * dS)S - *dS \wedge dS + dS \wedge *dS + Sd * dS$$
$$= -2(dS)^2 - 2(*dS)^2 + (d * dS)S + Sd * dS$$
$$= 2dS \wedge *dS + (d * dS)S + Sd * dS.$$

Now, together with $*Q - *A = \frac{1}{2}dS$ and $A = \frac{1}{4}(SdS + *dS)$, this implies

$$8d * Q = 8d * A = 2d(S * dS)$$
$$= 2dS \wedge *dS + 2Sd * dS$$
$$= -(d * dS)S + Sd * dS$$
$$= S(\underbrace{d * dS + S(d * dS)S}_{=2(d*dS)^T}).$$

We now consider the case where S is the mean curvature sphere of an immersed holomorphic curve. We decompose dS into the Hopf fields.

Lemma 6.

$$< dS \wedge *dS > = 4(< A \wedge *A > + < Q \wedge *Q >), \tag{6.6}$$
$$< dS \wedge SdS > = 4(< A \wedge *A > - < Q \wedge *Q >). \tag{6.7}$$

Proof. Recall from section 5.1

$$dS = 2(*Q - *A), \quad *dS = 2(A - Q), \quad SdS = 2(Q + A).$$

Further

$$*Q \wedge A = 0, \quad *A \wedge Q = 0$$

by type. Therefore

$$\begin{aligned} < dS \wedge *dS > &= 4 < (*Q - *A) \wedge (A - Q) > \\ &= -4 < *Q \wedge Q > -4 < *A \wedge A > \\ &= 4 < Q \wedge *Q > +4 < A \wedge *A >, \end{aligned}$$

and similarly for $< dS \wedge SdS >$.

Lemma 7. *Let V be a quaternionic vector space, $L \subset V$ a quaternionic line, $S, B \in \mathrm{End}(V)$ such that*

$$S^2 = -I, \quad SB = -BS, \quad \text{image } B \subset L.$$

Then

$$\mathrm{trace}_{\mathbb{R}} B^2 \leq 0,$$

with equality if and only if $B|_L = 0$.

Proof. We may assume $B \neq 0$. Then $L = BV$, and $SB = -BS$ implies $SL = L$. Let $\phi \in L \backslash \{0\}$, and

$$S\phi = \phi\lambda, \quad B\phi = \phi\mu.$$

Then $\lambda^2 = -1$, and $BS = -SB$ implies

$$\lambda\mu = -\mu\lambda.$$

Therefore μ is imaginary, too. It follows $B^2\phi = -|\mu|^2\phi$, and

$$\mathrm{trace}_{\mathbb{R}} B^2 = \mathrm{trace}_{\mathbb{R}} B^2|_L = -4|\mu|^2.$$

This can be applied to A or Q instead of B, since their rank is ≤ 1. We obtain

Lemma 8. *For an immersed holomorphic curve L we have*

$$< A \wedge *A > = \frac{1}{2} < A|_L \wedge *A|_L >, \tag{6.8}$$

and

$$< A \wedge *A > \geq 0, \quad < Q \wedge *Q > \geq 0. \tag{6.9}$$

In particular $E(S) \geq 0$.

Proof.

$$< A \wedge *A > = \frac{1}{8} \operatorname{trace}_{\mathbb{R}}(-A^2 - \underbrace{(*A)^2}_{=-ASSA=A^2}) = -\frac{1}{4} \operatorname{trace}_{\mathbb{R}} A^2.$$

Because $\dim L = \frac{1}{2} \dim H$ we similarly have

$$< A|_L \wedge *A|_L >= -\frac{1}{2} \operatorname{trace}_{\mathbb{R}} A|_L^2,$$

see section 5.3. Because $AH \subset L$, we have

$$\operatorname{trace}_{\mathbb{R}} A^2 = \operatorname{trace}_{\mathbb{R}} A|_L^2.$$

This proves (6.8). The positivity follows from Lemma 7.

Proposition 6. *1. The (alternating!) 2-form $\omega \in \Omega^2(\mathcal{Z})$ defined by*

$$\omega_S(X, Y) =< X, SY >, \quad for \ S \in \mathcal{Z}, \ X, Y \in T_S\mathcal{Z},$$

is closed.
  2. If $S : M \to \mathcal{Z}$, and $dS = 2(*Q - *A)$ as usual, see section 5.1 (5.3), then

$$S^*\omega = 2 < A \wedge *A > -2 < Q \wedge *Q >.$$

In particular,

$$\deg S := \frac{1}{\pi} \int_M < A \wedge *A > - < Q \wedge *Q >$$

is a topological invariant of S.

Remark 7. Since S maps the surface M into the 8-dimensional $\mathcal{Z}$, $\deg S$ certainly is not the mapping degree of S. But for immersed holomorphic curves it is the difference of two mapping degrees $\deg S = \deg N - \deg R$, where $N, R : M \to S^2$ are the left and right normal vector in affine coordinates, see chapter 7.

Proof. (i). We consider the 2-form on $\operatorname{End}(\mathbb{H}^2)$ defined by

$$\tilde{\omega}_S(X, Y) := \frac{1}{2}(< X, SY > - < Y, SX >).$$

Then $d_S\tilde{\omega}(X, Y, Z)$ is a linear combination of terms of the form

$$< Y, XZ >.$$

But if $X, Y, Z \in T_S\mathcal{Z}$, $S \in \mathcal{Z}$, we get

$$< Y, XZ > = - < S^2YXZ > = < SYXZS >$$
$$= < S^2YXZ > = - < Y, XZ >,$$

hence $< Y, XZ > = 0$. Therefore, if $\iota : \mathcal{Z} \to \text{End}(\mathbb{H}^2)$ is the inclusion,

$$d\omega = d\iota^*\tilde{\omega} = \iota^*d\tilde{\omega} = 0.$$

(ii). We have

$$
\begin{aligned}
S^*\omega(X,Y) &= < dS(X), SdS(Y) > \\
&= \frac{1}{2}(< dS(X)SdS(Y) > - < SdS(X)dS(Y) >) \\
&= \frac{1}{2}(< dS(X)SdS(Y) > - < dS(Y)SdS(X) >) \\
&= \frac{1}{2} < dS \wedge SdS > (X,Y),
\end{aligned}
$$

and Lemma 6 yields the formula.

The topological invariance under deformations of S follows from Stokes theorem: If $\tilde{S} : M \times [0,1] \to \mathcal{Z}$ deforms $S_0 : M \to \mathcal{Z}$ into S_1, then

$$
\begin{aligned}
0 &= \int_{M \times [0,1]} d\tilde{S}^*\omega \\
&= \int_{M \times 1} \tilde{S}^*\omega - \int_{M \times 0} \tilde{S}^*\omega \\
&= \int_M S_1^*\omega - \int_M S_0^*\omega.
\end{aligned}
$$

Remark 8. From

$$
\begin{aligned}
E(S) &= 4\int_M < A \wedge *A > + < Q \wedge *Q > \\
&= 8\int_M < A \wedge *A > + 4\underbrace{\int_M (< Q \wedge *Q > - < A \wedge *A >)}_{\text{topological invariant}}
\end{aligned}
$$

we see that for variational problems the energy functional can be replaced by the integral of $< A \wedge *A >$.

6.2 The Willmore Functional

Definition 8. *Let L be a compact immersed holomorphic curve in $\mathbb{H}P^1$ with Hopf field A. The Willmore functional of L is defined as*

$$W(L) := \frac{1}{\pi}\int_M < A \wedge *A > .$$

If we vary the immersion $L : M \to \mathbb{HP}^1$, it will in general not remain a holomorphic curve. On the other hand, any immersion induces a complex structure J on M such that with respect to this it is a holomorphic curve, see Proposition 3. Critical points of W with respect to such variations are called Willmore surfaces. If we consider only variations of L fixing the conformal structure of M they are called constrained Willmore surfaces, but we shall not treat this case here.

Example 19. For immersed surfaces in $\mathbb{R}^4$ we have

$$W(L) = \frac{1}{4\pi} \int_M (H^2 - K - K^\perp) |df|^2,$$

see section 7.3, Proposition 13.

Theorem 3 (Ejiri [2], Rigoli [12]). *An immersed holomorphic curve L is Willmore if and only if its mean curvature sphere S is harmonic.*

Proof. Let L_t be a variation, and S_t its mean curvature sphere. Note that for L_t to stay conformal the complex structure, i.e. the operator $*$, varies, too. The variation has a variational vector field $Y \in \Gamma(\text{Hom}(L, H/L))$ given by

$$Y\psi := \pi(\frac{d}{dt}\Big|_{t=0} \psi), \quad \psi_t \in \Gamma(L_t).$$

As usual, we abbreviate $\frac{d}{dt}|_{t=0}$ by a dot. Note that for $\psi \in \Gamma(L)$

$$\pi \dot{S}\psi = \pi(S\psi)\dot{} - \pi S\dot{\psi} = YS\psi - S\pi\dot{\psi} = (YS - SY)\psi. \tag{6.10}$$

We now compute the variation of the energy, which is as good as the Willmore functional as long as we vary L. By contrast, in the proof of Proposition 5 the conformal structure on M was fixed, and no L was involved.

$$\frac{d}{dt}\Big|_{t=0} E(S_t) = \frac{d}{dt}\Big|_{t=0} \int_M < dS_t \wedge *_t dS_t >$$

$$= \underbrace{\int_M < d\dot{S} \wedge *dS >}_{I} + \underbrace{\int_M < dS \wedge \dot{*}dS >}_{II} + \underbrace{\int_M < dS \wedge *d\dot{S} >}_{III}.$$

In general $< A \wedge *B > = < B \wedge *A >$, because $\text{trace}_{\mathbb{R}}(AB) = \text{trace}_{\mathbb{R}}(BA)$. Hence

$$III = I. \tag{6.11}$$

Next we claim

$$II = 0. \tag{6.12}$$

On TM let $\dot{J} = B$, i.e. $*\omega(X) =: \omega(BX)$. Then we have $BJ + JB = 0$, and

$$< dS \wedge *dS > (X, JX) = < dS(X)*dS(JX) > - < dS(JX)*dS(X) >$$
$$= < dS(X)dS(BJX) > - < dS(JX)dS(BX) >$$
$$= - < dS(X)dS(JBX) > - < dS(BX)dS(JX) > .$$

But S is conformal, see Proposition 4, therefore

$$< dS(X)dS(JX) >= 0 \text{ for all } X.$$

Differentiation with respect to X yields

$$< dS(X)dS(JY) > + < dS(Y)dS(JX) >= 0$$

for all X, Y. Using this with $Y = BX$ we get (6.12).

Now, we compute the integral I.

$$I = -\int_M < \dot{S}, d * dS >$$
$$\underset{(6.5)}{=} 4 \int_M < \dot{S}, Sd * Q >$$
$$= \frac{1}{2} \int_M \text{trace}_{\mathbb{R}}(\dot{S}Sd * Q).$$

We shall show in the following lemma that

$$\text{image } d * Q \subset L \subset \ker d * Q.$$

Therefore we can consider $d * Q$ as a 2-form

$$d * Q \in \Omega^2(\text{Hom}(H/L, L)),$$

and continue

$$I = \frac{1}{2} \int_M \text{trace}_{\mathbb{R}}(\dot{S}Sd * Q : H \to H)$$
$$= \frac{1}{2} \int_M \text{trace}_{\mathbb{R}}(\pi \dot{S}Sd * Q : H/L \to H/L)$$
$$= \frac{1}{2} \int_M \text{trace}_{\mathbb{R}}(\pi \dot{S}|_L Sd * Q : H/L \to H/L)$$
$$\underset{(6.10)}{=} \frac{1}{2} \int_M \text{trace}_{\mathbb{R}}((YS - SY)(Sd * Q) : H/L \to H/L)$$
$$= -\frac{1}{2} \int_M \text{trace}_{\mathbb{R}}(Yd * Q) - \frac{1}{2} \int_M \text{trace}_{\mathbb{R}}(SYSd * Q).$$

Now $d * Q$ is tangential by (6.5), and hence anti-commutes with S. Thus

$$I = -\frac{1}{2} \int_M \mathrm{trace}_{\mathbb{R}}(Y d * Q) + \frac{1}{2} \int_M \mathrm{trace}_{\mathbb{R}}(SY d * QS)$$

$$= - \int_M \mathrm{trace}_{\mathbb{R}}(Y d * Q)$$

$$= -8 \int_M < Y, d * Q >$$

We therefore showed

$$\frac{d}{dt}\Big|_{t=0} E(S_t) = -8 \int_M < Y, d * Q > .$$

Since $Y \in \Omega^2(\mathrm{Hom}(H/L, L))$, this vanishes for all variational vector fields Y if and only if

$$d * Q = 0.$$

In the proof we made use of the following

Lemma 9.

$$\mathrm{image}\, d * Q \subset L \subset \ker d * Q.$$

Proof. For $\psi \in \Gamma(L)$

$$0 = d(*Q\psi) = (d * Q)\psi - *Q \wedge d\psi = (d * Q)\psi - *Q \wedge \delta\psi,$$

because $Q|_L = 0$. But $*Q$ is right K, and δ is left K. Hence, by type,

$$(d * Q)\psi = *Q \wedge \delta\psi = 0.$$

This shows the right hand inclusion. Also,

$$\pi(d * Q)(X, JX) = \pi(d * A)(X, JX)$$
$$= \pi(X \cdot (*A(JX)) - JX \cdot (*A(X)) - \underbrace{*A([X, JX])}_{L-\mathrm{valued}})$$
$$= \delta(X) * A(JX) - \delta(JX) * A(X)$$
$$= -\delta(X)A(X) - \delta(X)SSA(X)$$
$$= 0.$$

7 Metric and Affine Conformal Geometry

We consider the metric extrinsic geometry of $f : M \to \mathbb{R}^4$ in relation to the quantities associated to

$$L := \begin{bmatrix} f \\ 1 \end{bmatrix} : M \to \mathbb{H}P^1.$$

For brevity we write $< ., . >$ instead of $< ., . >_\mathbb{R}$.

7.1 Surfaces in Euclidean Space

Let N, R denote the left and right normal vector of $f : M \to \mathbb{H}$, i.e.

$$*df = N df = -df R.$$

Proposition 7. *The second fundamental form* $II(X,Y) = (X \cdot df(Y))^\perp$ *of f is given by*

$$II(X,Y) = \frac{1}{2}(*df(Y)dR(X) - dN(X) * df(Y)). \tag{7.1}$$

Proof. We know from Lemma 2 that $v \mapsto N(x)vR(x)$ is an involution with the tangent space as its fixed point set:

$$N df(Y)R = df(Y) \tag{7.2}$$

Its (-1)-eigenspace is the normal space, so we need to compute

$$II(X,Y) = \frac{1}{2}(X \cdot df(Y) - NX \cdot df(Y)R).$$

But differentiation of (7.2) yields

$$dN(X)df(Y)R + NX \cdot df(Y)R + N df(Y)dR(X) = X \cdot df(Y),$$

or

$$X \cdot df(Y) - NX \cdot df(Y)R = dN(X)df(Y)R + N df(Y)dR(X)$$
$$= -dN(X) * df(Y) + *df(Y)dR(X).$$

Proposition 8. *The mean curvature vector* $\mathcal{H} = \frac{1}{2}\operatorname{trace} II$ *is given by*

$$\mathcal{H}df = \frac{1}{2}(*dR + RdR), \quad df\bar{\mathcal{H}} = -\frac{1}{2}(*dN + NdN). \tag{7.3}$$

Proof. By definition of the trace,

$$4\mathcal{H}|df|^2 = *df\,dR - dN * df - df * dR + *dN\,df \tag{7.4}$$
$$= -df(*dR + RdR) + (*dN + NdN)df, \tag{7.5}$$

but

$$(*dN + NdN)df = *dN\,df - dN * df = -dN \wedge df = -d(N\,df)$$
$$= -df \wedge dR = -df(*dR + RdR).$$

If follows that

$$2\mathcal{H}|df|^2 = -df(*dR + RdR),$$

and

$$2\bar{\mathcal{H}}df\,\overline{df} = -(- * dR + dRR)\overline{df} = (*dR + RdR)\overline{df}.$$

Similarly for N.

Proposition 9. *Let K denote the Gaussian curvature of $(M, f^* < .,. >_{\mathbb{R}})$ and let $K^{\perp}$ denote the normal curvature of f defined by*

$$K^{\perp} :=< R^{\perp}(X, JX)\xi, N\xi >_{\mathbb{R}},$$

where $X \in T_pM$, and $\xi \in \perp_p M$ are unit vectors. Then

$$K|df|^2 = \frac{1}{2}(< *dR, RdR > + < *dN, NdN >) \tag{7.6}$$

$$K^{\perp}|df|^2 = \frac{1}{2}(< *dR, RdR > - < *dN, NdN >) \tag{7.7}$$

Proof.

$$K|df|^4(X) =< II(X, X), II(JX, JX) > -|II(X, JX)|^2.$$

Therefore

$$4K|df|^4 = < *df\,dR - dN * df, -df * dR + *dN\,df >$$
$$- < *df * dR - *dN * df, -df\,dR + dN\,df >$$
$$= < N(df\,dR + dN\,df), -df * dR + *dN\,df >$$
$$- < N(df * dR + *dN\,df), -df\,dR + dN\,df >$$
$$= - < df\,dR + dN\,df, N(-df * dR + *dN\,df) >$$
$$< df * dR + *dN\,df, N(-df\,dR + dN\,df) >$$
$$= - < df\,dR + dN\,df, df\,R * dR + N * dN\,df >$$
$$+ < df * dR + *dN\,df, df\,R\,dR + N\,dN\,df >$$
$$= - < df\,dR, df\,R * dR > - < df\,dR, N * dN\,df >$$
$$- < dN\,df, df\,R * dR > - < dN\,df, N * dN\,df >$$
$$+ < df * dR, df\,R\,dR > + < df * dR, N\,dN\,df >$$
$$+ < *dN\,df, df\,R\,dR > + < *dN\,df, N\,dN\,df >$$
$$= - |df|^2 < dR, R * dR > - < df\,dR, N * dN\,df >$$
$$+ < dN\,df, N\,df * dR > - |df|^2 < dN, N * dN >$$
$$+ |df|^2 < *dR, R\,dR > + < df * dR, N\,dN\,df >$$
$$- < *dN\,df, N\,df\,dR > + |df|^2 < *dN, N\,dN >$$
$$= - |df|^2 (< dR, R * dR > + < dN, N * dN >$$
$$- < *dR, R\,dR > - < *dN, N\,dN >)$$
$$= - 2|df|^2 (< dR, R * dR > + < dN, N * dN >).$$

This proves the formula for K. Using (7.1) and the Ricci equation

$$K^\perp = < N\,II(X, JX), II(X, X) - II(JX, JX) >,$$

we find, after a similar computation,

$$4K^\perp |df|^2 = < *dR - R\,dR, R\,dR > - < *dN - N\,dN, N\,dN >$$
$$+ < df(*dR - R\,dR), N\,dN\,df > - < (*dN - N\,dN)df, df\,R\,dR > .$$

On this we use (7.5) to obtain (7.7).

As a corollary we have

Proposition 10. *The pull-back of the 2-sphere area under R is given by*

$$R^* dA = < *dR, R\,dR > .$$

Integrating this for compact M yields

$$\frac{1}{4\pi} \int_M K|df|^2 = \frac{1}{2}(\deg R + \deg N).$$

In 3-space $(R = N)$ this is a version of the Gauss-Bonnet theorem.

Proposition 11. *We obtain*

$$(|\mathcal{H}|^2 - K - K^\perp)|df|^2 = \frac{1}{4}|*dR - RdR|^2$$

In particular, if $f : M \to \operatorname{Im}\mathbb{H} = \mathbb{R}^3$ then $K^\perp = 0$, and the classical Willmore integrand is given by

$$(|\mathcal{H}|^2 - K)|df|^2 = \frac{1}{4}|*dR - RdR|^2. \tag{7.8}$$

Proof. Equations (7.3), (7.6), (7.7) give

$$\begin{aligned}
(|\mathcal{H}|^2 - K - K^\perp)|df|^2 &= \frac{1}{4}|*dR + RdR|^2 - <*dR, RdR> \\
&= \frac{1}{4}|*dR|^2 + \frac{1}{4}|RdR|^2 - \frac{1}{2}<*dR, RdR> \\
&= \frac{1}{4}|*dR - RdR|^2.
\end{aligned}$$

7.2 The Mean Curvature Sphere in Affine Coordinates

We now discuss the characteristic properties of S in affine coordinates. We describe S relative to the frame $\begin{pmatrix} 1 \\ 0 \end{pmatrix}, \begin{pmatrix} f \\ 1 \end{pmatrix}$, i.e. we write $S = GMG^{-1}$, where

$$G = \begin{pmatrix} 1 & f \\ 0 & 1 \end{pmatrix}.$$

First, $SL \subset L$ is equivalent to $S : \mathbb{H}^2 \to \mathbb{H}^2$ having the following matrix representation:

$$S = \begin{pmatrix} 1 & f \\ 0 & 1 \end{pmatrix} \begin{pmatrix} N & 0 \\ -H & -R \end{pmatrix} \begin{pmatrix} 1 & -f \\ 0 & 1 \end{pmatrix} \tag{7.9}$$

where $N, R, H : M \to \mathbb{H}$. From $S^2 = -I$

$$N^2 = -1 = R^2, \quad RH = HN. \tag{7.10}$$

The choice of symbols is deliberate: N and R turn out to be the left and right normal vectors of f, while H is closely related to its mean curvature vector $\mathcal{H}$.

The bundle L has the nowhere vanishing section $\begin{pmatrix} f \\ 1 \end{pmatrix} \in \Gamma(L)$. Using this section, we compute

$$*\delta \begin{pmatrix} f \\ 1 \end{pmatrix} = \pi \begin{pmatrix} *df \\ 0 \end{pmatrix},$$

$$\delta S \begin{pmatrix} f \\ 1 \end{pmatrix} = \pi d(S \begin{pmatrix} f \\ 1 \end{pmatrix}) = \pi d(\begin{pmatrix} f \\ 1 \end{pmatrix}(-R)) = \pi(\begin{pmatrix} -df\,R \\ 0 \end{pmatrix} + \begin{pmatrix} f \\ 1 \end{pmatrix}(-dR)) = \pi \begin{pmatrix} -df\,R \\ 0 \end{pmatrix},$$

$$S\delta \begin{pmatrix} f \\ 1 \end{pmatrix} = \pi S d \begin{pmatrix} f \\ 1 \end{pmatrix} = \pi(\begin{pmatrix} N df \\ 0 \end{pmatrix} + \begin{pmatrix} f \\ 1 \end{pmatrix}(-Hdf)) = \pi \begin{pmatrix} N df \\ 0 \end{pmatrix}.$$

Therefore $*\delta = S\delta = \delta S$ is equivalent to

$$*df = N df = -df\,R,$$

and we have identified N and R.

For the computation of the Hopf fields, we need dS. This is a straightforward but lengthy computation, somewhat simplified by the fact that $G dG = dG = G^{-1} dG$. We skip the details and give the result:

$$dS = G \begin{pmatrix} -df\,H + dN & -df\,R - N df \\ -dH & -dR + Hdf \end{pmatrix} G^{-1},$$

$$S dS = G \begin{pmatrix} -N df\,H + N dN & 0 \\ Hdf\,H + RdH - HdN & Hdf\,R + RdR \end{pmatrix} G^{-1}.$$

From this we obtain

$$4Q = S dS - *dS$$
$$= G \begin{pmatrix} N dN - *dN & 0 \\ *dH + Hdf\,H + RdH - HdN & 2Hdf\,R + RdR + *dR \end{pmatrix} G^{-1}$$

$$4A = S dS + *dS$$
$$= G \begin{pmatrix} N dN + *dN - 2N df\,H & 0 \\ - *dH + Hdf\,H + RdH - HdN & RdR - *dR \end{pmatrix} G^{-1}.$$

The condition $Q|_L = 0$, and the corresponding $AH \subset L$, which we have not used so far, have the following equivalents:

$$2Hdf = dR - R * dR, \tag{7.11}$$
$$2df\,H = dN - N * dN. \tag{7.12}$$

Together with equations (7.3) we find

$$2Hdf = dR - R * dR = -R(*dR + RdR) = -2R\bar{\mathcal{H}}df,$$
$$2df\,H = dN - N * dN = -N(*dN + N dN) = 2N df\,\bar{\mathcal{H}} = -2df\,R\bar{\mathcal{H}},$$

and therefore

$$H = -\bar{\mathcal{H}}N = -R\bar{\mathcal{H}}. \tag{7.13}$$

Remark 9. Given an immersed holomorphic curve $L = \begin{bmatrix} f \\ 1 \end{bmatrix}$, the mean curvature vector of f at $x \in M$ is determined by S_x. On the other hand, S_x is the mean curvature sphere of S_x, see Example 17. Therefore S_x and f have, in fact, the same mean curvature vector at x, justifying the name *mean curvature sphere.*

Equations (7.11), (7.12) simplify the coordinate expressions for the Hopf fields, which we now write as follows

Proposition 12.

$$4 * Q = G \begin{pmatrix} dN + N * dN & 0 \\ -2dH + w & 0 \end{pmatrix} G^{-1}, \tag{7.14}$$

$$4 * A = G \begin{pmatrix} 0 & 0 \\ w \, dR + R * dR & \end{pmatrix} G^{-1}, \tag{7.15}$$

where $G = \begin{pmatrix} 1 & f \\ 0 & 1 \end{pmatrix}$, and $w = dH + H * df H + R * dH - H * dN$.
Using (7.12) we can rewrite

$$w = dH + R * dH + \frac{1}{2} H(NdN - *dN).$$

Proof. We only have to consider the reformulation of w. But

$$H * df H - H * dN = \frac{1}{2} H * (dN - N * dN) - H * dN$$

$$= -\frac{1}{2} H * (dN + N * dN) = \frac{1}{2} H(NdN - *dN).$$

7.3 The Willmore Condition in Affine Coordinates

We use the notations of the previous Proposition 12, and in addition abbreviate

$$v = dR + R * dR.$$

Note that

$$\bar{v} = -dR + *dRR = -dR - R * dR = -v.$$

Proposition 13. *The Willmore integrand is given by*

$$< A \wedge *A > = \frac{1}{16} |RdR - *dR|^2 = \frac{1}{4} (|\mathcal{H}|^2 - K - K^\perp)|df|^2.$$

For $f : M \to \mathbb{R}^3$, this is the classical integrand

$$< A \wedge *A > = \frac{1}{4} (|\mathcal{H}|^2 - K)|df|^2.$$

Proof.

$$< A \wedge *A > = \frac{1}{8} \operatorname{trace}_{\mathbb{R}}(-A^2 - (*A)^2) = -\frac{1}{4} \operatorname{trace}_{\mathbb{R}}(A^2)$$

$$= -\frac{1}{4} 4 \operatorname{Re}(\frac{1}{4}v)^2 = \frac{1}{16}|v|^2 = \frac{1}{16}|dR + R*dR|^2 = \frac{1}{16}|RdR - *dR|^2.$$

Now see Proposition 11 and, for the second equality, (7.8).

We now express the Euler-Lagrange equation $d * A = 0$ for Willmore surfaces in affine coordinates. If we write $4 * A = GMG^{-1}$, then

$$4d * A = G(G^{-1}dG \wedge M + dM + M \wedge G^{-1}dG)G^{-1},$$

and again using $G^{-1}dG = dG$ we easily find

$$4d * A = G \begin{pmatrix} df \wedge w & df \wedge v \\ dw & dv + w \wedge df \end{pmatrix} G^{-1}.$$

Most entries of this matrix vanish:

Proposition 14. *We have*

$$df \wedge w = 0 \tag{7.16}$$
$$df \wedge v = 0 \tag{7.17}$$
$$dv + w \wedge df = -(2dH - w) \wedge df = 0. \tag{7.18}$$

Proof. We have

$$df \wedge w = df \wedge dH + df \wedge R * dH + \frac{1}{2}df \wedge H(NdN - *dN)$$

$$= df \wedge dH + df R \wedge *dH + \frac{1}{2}df H \wedge (NdN - *dN)$$

$$= \underbrace{df \wedge dH - *df \wedge *dH}_{=0} + \frac{1}{2}df H \wedge (NdN - *dN),$$

but

$$*df H = df(-R)H = -df HN$$
$$*(NdN - *dN) = (N * dN - N^2 dN) = -N(NdN - *dN).$$

Hence, by type, the second term vanishes as well, and we get (7.16).
  A similar, but simpler, computation shows (7.17)
  Next, using (7.11), we consider

$$dv + w \wedge df = d(dR + R * dR) + w \wedge df$$
$$= d(-2Hdf) + w \wedge df$$
$$= (-2dH + w) \wedge df$$
$$= (\underbrace{-dH + R * dH}_{=:\alpha} + \underbrace{\frac{1}{2}H(NdN - *dN)}_{\beta}) \wedge df.$$

Again we show $*\alpha = \alpha N, *\beta = \beta N$. Then (7.18) will follow by type.
  Clearly

$$*(NdN - *dN) = N * dN + NdNN = (NdN - *dN)N,$$

showing $*\beta = \beta N$. Further

$$
\begin{aligned}
*\alpha - \alpha N &= - * dH - RdH + dHN - R(*dH)N \\
&= - * dH - d(RH) + (dR)H + d(\underbrace{HN}_{=RH}) - HdN - R * (d(\underbrace{HN}_{=RH}) - HdN) \\
&= +R^2 * dH + (dR)H - HdN - R * ((dR)H + RdH - HdN) \\
&= (dR)H - HdN - R * (dR)H + RH * dN) \\
&= (dR - R * dR)H - H(dN - N * dN) \\
&= 2H\,df\,H - H(2df\,H) \\
&= 0.
\end{aligned}
$$

As a corollary we get:

Proposition 15.

$$d * A = \frac{1}{4}G \begin{pmatrix} 0 & 0 \\ dw & 0 \end{pmatrix} G^{-1} = \frac{1}{4}\begin{pmatrix} f\,dw & -f\,dw\,f \\ dw & -dw\,f \end{pmatrix}.$$

with $w = dH + R * dH + \frac{1}{2}H(NdN - *dN)$.
  Therefore f is Willmore if and only if $dw = 0$.

Example 20 (Willmore Cylinder). Let $\gamma : \mathbb{R} \to \mathrm{Im}\,\mathbb{H}$ be a unit-speed curve, and $f : \mathbb{R}^2 \to \mathbb{H}$ the cylinder defined by

$$f(s,t) = \gamma(s) + t$$

with the conformal structure $J\frac{\partial}{\partial s} = \frac{\partial}{\partial t}$. Then using Proposition 15, we obtain, after some computation, that f is (non-compact) Willmore, if and only if

$$\frac{1}{2}\kappa^3 + \kappa'' - \kappa\tau^2 = 0, \quad (\kappa^2\tau)' = 0.$$

This is exactly the condition that γ be a free elastic curve.

8 Twistor Projections

8.1 Twistor Projections

Let $E \subset H := M \times \mathbb{H}^2 = M \times \mathbb{C}^4$ be a *complex* (not a quaternionic) line subbundle over a Riemann surface M with complex structure J_E induced from right multiplication by i on $\mathbb{H}^2$.

We define $\delta_E \in \Omega^1(\mathrm{Hom}(E, H/E))$ by

$$\delta_E \phi := \pi_E d\phi, \quad \phi \in \Gamma(E),$$

where $\pi_E : H \to H/E$ is the projection.

Definition 9. *E is called a* holomorphic curve *in* $\mathbb{C}P^3$, *if*

$$*\delta_E = \delta_E J_E.$$

This is equivalent to the fact that the holomorphic structure

$$d''\psi = \frac{1}{2}(d\psi + i * d\psi) \tag{8.1}$$

of H maps $\Gamma(E)$ into itself, and hence induces a holomorphic structure on the complex line bundle E.

A complex line bundle $E \subset H$ induces a quaternionic line bundle

$$L = E\mathbb{H} = E \oplus Ej \subset H.$$

The complex structure J_E admits a unique extension to the structure of a complex quaternionic bundle (L, J), namely right-multiplication by $(-i)$ on Ej. Conversely, a complex quaternionic line bundle $(L, J) \subset H$ induces a complex line bundle

$$E := \{\phi \in L \mid J\phi = \phi i\}.$$

Definition 10. *We call (L, J) the twistor projection of E, and E the twistor lift of (L, J).*

Remark 10. As in the quaternionic case, any map $f : M \to \mathbb{C}P^3$ induces a complex line bundle E, where the fibre over p is $f(p)$, and vice versa. Holomorphic curves as defined above correspond to holomorphic curves in the sense of complex analysis. The correspondence between E and (L, J) is mediated by the Penrose twistor projection $\mathbb{C}P^3 \to \mathbb{H}P^1$.

Theorem 4. *Let $E \subset H$ be a a complex line subbundle over a Riemann surface M, and (L, J) its twistor projection.*

1. *Then (L,J) is a holomorphic curve, i.e.*

$$*\delta_L = \delta_L J, \tag{8.2}$$

   if and only if

$$\frac{1}{2}(\delta_E + *\delta_E J_E) \in \Omega^1(\mathrm{Hom}(E, L/E)) \subset \Omega^1(\mathrm{Hom}(E, H/E)).$$

   In this case we have a differential operator

$$\tilde{D} : \Gamma(L) \to \Omega^1(L), \psi \mapsto \tilde{D}\psi := \frac{1}{2}(d\psi + *d(J\psi))$$

   Its $(1,0)$-part is given by

$$A_L := \frac{1}{2}(\tilde{D} + J\tilde{D}J) \in \Gamma(K \, \mathrm{End}_-(L)). \tag{8.3}$$

2. *If (L, J) is a holomorphic curve then*

$$\frac{1}{2}(\delta_E + *\delta_E J_E) = \pi_E A_L|_E.$$

   Moreover,

$$\frac{1}{2}(\delta_E + *\delta_E J_E) = 0 \iff A_L = 0.$$

   In other words: The twistor projections of holomorphic curves in $\mathbb{C}P^3$ are exactly the holomorphic curves in $\mathbb{H}P^1$ with $A_L = 0$.
3. *Let L be an immersed holomorphic curve with mean curvature sphere congruence $S \in \Gamma(\mathrm{End}_-(H))$, and $J = S|_L$. Then*

$$A = \frac{1}{4}(SdS + *dS) \in \Gamma(\bar{K} \, \mathrm{End}_-(H))$$

   satisfies

$$A|_L = A_L.$$

Proof. (i). If (L, J) is a holomorphic curve then, for any $\psi \in \Gamma(L)$,

$$\frac{1}{2}\pi_L(d\psi + *d(J\psi)) = 0.$$

But then

$$\frac{1}{2}(d\psi + *d(J\psi)) \in \Omega^1(L)$$

a fortiori for all $\psi = \phi \in \Gamma(E)$. It follows

$$\frac{1}{2}\pi_E(d\phi + *d(J_E\phi)) \in \Omega^1(L/E).$$

Conversely, $\frac{1}{2}\pi_E(d\phi + *d(J_E\phi)) \in \Omega^1(L/E)$ for $\phi \in \Gamma(E)$ implies

$$\frac{1}{2}(d\phi + *d(J_E\phi)) \in \Omega^1(L),$$

and therefore

$$*\delta_L|_E = \delta_L J|_E.$$

Again for $\phi \in \Gamma(E)$

$$\frac{1}{2}(d(\phi j) + *d(J\phi j)) = \frac{1}{2}((d\phi)j + *d(J\phi)j)) = \frac{1}{2}\underbrace{(d\phi + *d(J_E\phi))}_{\in \Omega^1(L)}j \in \Omega^1(L).$$

This shows

$$*\delta_L = \delta_L J.$$

By the preceding, $\tilde{D}$ maps into $\Omega^1(L)$. Its $(1,0)$-part is

$$\frac{1}{2}(\tilde{D} - J * \tilde{D}),$$

but for $\psi \in \Gamma(L)$

$$*\tilde{D}\psi = \frac{1}{2}(*d\psi - d(J\psi)) = -\tilde{D}J\psi.$$

This proves (8.3).

(ii). For $\psi \in \Gamma(L)$ we have

$$A_L\psi = \frac{1}{4}(d\psi + *d(J\psi) + J(dJ\psi - *d\psi)) \tag{8.4}$$

But for $\phi \in \Gamma(E)$ we have $J(dJ\phi - *d\phi) = J(d\phi + *d\phi i)i$, and hence

$$A_L\phi = \frac{1}{4}((d\phi + *d(J\phi)) + J(d\phi + *d(J\phi))i).$$

By assumption $\frac{1}{2}(d\phi + *d(J\phi))$ has values in $L = E \oplus Ej$, and $A_L\phi$ is its Ej-component, namely the component in the $(-i)$-eigenspace of $J|_L$. In particular,

$$\pi_E A_L\phi = \pi_E \frac{1}{2}(d\phi + *d(J\phi)) = \frac{1}{2}(\delta_E + *\delta_E J)\phi,$$

and $\pi_E(A_L\phi) = 0$ if and only if $A_L\phi = 0$. Since $A_L|_E$ determines A_L by linearity, $\frac{1}{2}\pi_E(d\psi + *d(\psi i)) = 0 \iff A_L = 0$.

(iii). For $\psi \in \Gamma(L)$

$$\begin{aligned}
A\psi &= \frac{1}{4}(SdS + *dS)\psi \\
&= \frac{1}{4}(S(d(S\psi) - Sd\psi) + *d(S\psi) - *Sd\psi) \\
&= \frac{1}{4}(S(d(S\psi) - *d\psi) + *d(S\psi) + d\psi).
\end{aligned}$$

Comparison with (8.4) shows $A|_L = A_L$.

In view of lemma 8 this implies the following

Corollary 1. *The twistor projections of holomorphic curves in $\mathbb{C}P^3$ are exactly the holomorphic curves in $\mathbb{H}P^1$ with vanishing Willmore functional.*

8.2 Super-Conformal Immersions

Given a surface conformally immersed into $\mathbb{R}^4$, the image of a tangential circle under the quadratic second fundamental form is (a double cover of) an ellipse in the normal space, centered at the mean curvature vector, the so-called *curvature ellipse*. The surface is called *super-conformal* if this ellipse is a circle.

If N and R are the left and right normal vector of f, then according to Proposition 7 we have

$$II(X,Y) = \frac{1}{2}(*df(Y)dR(X) - dN(X) * df(Y)),$$

and therefore

$$II(\cos\theta X + \sin\theta JX, \cos\theta X + \sin\theta JX)$$

$$=\frac{1}{2}(*df(\cos\theta X + \sin\theta JX)dR(\cos\theta X + \sin\theta JX)$$
$$- dN(\cos\theta X + \sin\theta JX) * df(\cos\theta X + \sin\theta JX))$$

$$=\frac{1}{2}(df(\cos\theta JX - \sin\theta X)dR(\cos\theta X + \sin\theta JX)$$
$$- dN(\cos\theta X + \sin\theta JX)df(\cos\theta JX - \sin\theta X))$$

$$=\frac{1}{2}(\cos^2\theta(df(JX)dR(X) - dN(X)df(JX))$$
$$- \sin^2\theta(df(X)dR(JX) - dN(JX)df(X))$$
$$+ \cos\theta\sin\theta(df(JX)dR(JX) - df(X)dR(X)$$
$$+ dN(X)df(X) - dN(JX)df(JX)).$$

Using $\cos^2\theta = \frac{1}{2}(1 + \cos 2\theta)$, $\sin^2\theta = \frac{1}{2}(1 - \cos 2\theta)$ we get

$$II(\cos\theta X + \sin\theta JX, \cos\theta X + \sin\theta JX)$$

$$=\frac{1}{4}(\underbrace{*df(X)dR(X) - dN(X)*df(X)}_{=2II(X,X)} + \underbrace{*df(JX)dR(JX) - dN(JX)*df(JX)}_{=2II(JX,JX)})$$

$$+\frac{1}{4}\cos 2\theta(df(JX)dR(X) - dN(X)df(JX) + df(X)dR(JX) - dN(JX)df(X))$$

$$+\frac{1}{4}\sin 2\theta(df(JX)dR(JX) - df(X)dR(X) + dN(X)df(X) - dN(JX)df(JX))$$

$$=\mathcal{H}|df(X)|^2$$

$$+\frac{1}{4}\cos 2\theta(\underbrace{df(X)(*dR(X) - RdR(X))}_{=:a} - \underbrace{(*dN(X) - NdN(X))df(X)}_{=:b})$$

$$+\frac{1}{4}\sin 2\theta N(a + b).$$

This is a circle if and only if $a - b$ and $N(a + b)$ are orthogonal and have same length. This is clearly the case if $a = 0$ or $b = 0$, but these are in fact the only possibilities: Assume that there exists $P \in \mathbb{H}, P^2 = -1$ with

$$N(a + b) = P(a - b), \tag{8.5}$$

and note that

$$Na = aR, Nb = bR.$$

We multiply (8.5) by N from the left or by R from the right to obtain

$$-(a + b) = NP(a - b), \qquad -(a + b) = PN(a - b)$$

respectively. Therefore $(PN - NP)(a - b) = 0$, which implies $P = \pm N$, and hence $a = 0$ or $b = 0$, or $a - b = 0$. But then also $a + b = 0$, whence $a = b = 0$.

It follows that the immersion is super-conformal if and only if

$$*dR(X) - RdR(X) = 0, \text{ or } * dN(X) - NdN(X) = 0.$$

By the preceding argument, this holds for a particular choice of X, but then it obviously follows for all X.

We mention that $f \to \bar{f}$ exchanges N and R, hence f is super-conformal, if and only if $*dR - RdR = 0$ for f or for $\bar{f}$. In view of proposition 12, this is equivalent to $A|_L = 0$, and by Theorem 4 we obtain:

Theorem 5. *A conformally immersed Riemann surface $f : M \to \mathbb{H} = \mathbb{R}^4$ is super-conformal if and only if* $\begin{bmatrix} f \\ 1 \end{bmatrix} : M \to \mathbb{HP}^1$ *or* $\begin{bmatrix} \bar{f} \\ 1 \end{bmatrix} : M \to \mathbb{HP}^1$ *is the twistor projection of a holomorphic curve in* $\mathbb{CP}^3$.

9 Bäcklund Transforms of Willmore Surfaces

In this section we shall describe a method to construct new Willmore surfaces from a given one. The construction depends on the choice of a point ∞, and therefore generously offers a 4-parameter family of such transformations. On the other hand, the necessary computations are not invariant, and therefore ought to be done in affine coordinates.

The transformation theory is essentially local: This fact will be hidden in the assumption that the transforms are again immersions. We shall also ignore period problems.

9.1 Bäcklund Transforms

Let $f : M \to \mathbb{H}$ be a Willmore surface with N, R, H, and

$$w = dH + H * df H + R * dH - H * dN.$$

Then

$$dw = 0,$$

and hence we can integrate it. Assume that $g : M \to \mathbb{H}$ is an immersion with

$$dg = \frac{1}{2}w. \tag{9.1}$$

(Note that the integral of $w/2$ may have periods, so in general g is defined only on a covering of M. We ignore this problem.)

We want to show that g is again a Willmore surface called a *Bäcklund transform* of f. Using this name, we refer to the fact that in a given category of surfaces we construct new examples from old ones by solving an ODE (9.1), similar to the classical Bäcklund transforms of K-surfaces, see Tenenblat [13].

We denote the symbols associated to g by a subscript $(.)_g$, and want to prove $dw_g = 0$. The computation of w_g can be done under the weaker assumption (9.2), which holds in the case above, see Proposition 14.

Proposition 16. *Let $f, g : M \to \mathbb{H}$ be immersions such that*

$$df \wedge dg = 0. \tag{9.2}$$

Then f and g induce the same conformal structure on M, and

$$N_g = -R, \tag{9.3}$$

$$dg(2dH_g - w_g) = -wdf. \tag{9.4}$$

Proof. Define $*$ using the conformal structure induced by f. Then

$$0 = df \wedge dg = df * dg - df(-R)dg,$$

which implies $*dg = -Rdg$. Hence g is conformal, too, and $N_g = -R$.

For the next computations recall the equations (7.10), and (7.11), (7.12):

$$HN = RH,$$

$$2dfH = dN - N * dN, \quad 2Hdf = dR - R * dR,$$

$$w = dH + H * dfH + R * dH - H * dN.$$

Then

$$\begin{aligned} Rw &= RdH + RH * dfH - *dH - RH * dN \\ &= RdH + HN * dfH - *dH - HN * dN \\ &= RdH - HdfH - *dH - H(N * dN - dN) - HdN \\ &= RdH - HdN + HdfH - *dH. \end{aligned} \tag{9.5}$$

With $dRH + RdH = dHN + HdN$ this becomes

$$Rw = dHN - *dH - dRH + HdfH. \tag{9.6}$$

Next

$$2dgH_g = dN_g - N_g * dN_g = -dR - R * dR.$$

Therefore

$$-dg \wedge dH_g = \frac{1}{2}d(-dR - R * dR) = \frac{1}{2}d(dR - R * dR) = dH \wedge df,$$

or

$$dg(*dH_g + R_g dH_g) = -(dHN - *dH)df. \tag{9.7}$$

We now use (9.5) and (9.7) to compute

$$\begin{aligned} &N_g dg(2dH_g - w_g) \\ &= -dgR_g(2dH_g - w_g) \\ &= dg(-2R_g dH_g + R_g dH_g - H_g dN_g + H_g dgH_g - *dH_g) \\ &= -dg(R_g dH_g + *dH_g) + dgH_g(dgH_g - dN_g) \\ &= (dHN - *dH)df + dgH_g(dgH_g - dN_g) \\ &= (dHN - *dH)df + \frac{1}{4}(dN_g - N_g * dN_g)((dN_g - N_g * dN_g) - 2dN_g) \\ &= (dHN - *dH)df - \frac{1}{4}(dR + R * dR)(dR - R * dR). \end{aligned}$$

Similarly, using (9.6),

$$
\begin{aligned}
-N_g w df &= R w df \\
&= (dHN - *dH)df - (dR - Hdf)Hdf \\
&= (dHN - *dH)df - \frac{1}{4}(2dR - dR + R*dR)(dR - R*dR) \\
&= (dHN - *dH)df - \frac{1}{4}(dR + R*dR)(dR - R*dR).
\end{aligned}
$$

Comparison yields (9.4).

If f is Willmore, and g is defined by (9.1), then

$$
dg(2df + 2dH_g - w_g) = 2dgdf + dg(2dH_g - w_g) = (2dg - w)df = 0.
$$

Hence

$$
w_g = 2d(f + H_g), \tag{9.8}
$$

and g is Willmore, too.

Now assume that $h := g - H$ is again an immersion. Then, by Proposition 14,

$$
2dh \wedge df = (2dg - 2dH) \wedge df = (w - 2dH) \wedge df = 0.
$$

Proposition 16 applied to (h, f) instead of (f, g) then says

$$
-w_h dh = df(2dH - w) = df(2dH - 2dg) = -2df dh.
$$

We find $w_h = 2df$, whence h is again a Willmore surface. We call g a *forward*, and h a *backward Bäcklund transform* of f. h can be obtained without reference to g by integrating $d(g - H) = \frac{1}{2}w - dH$.

Note that f is a forward Bäcklund transform of h because $df = \frac{1}{2}w_h$, and is also a backward transform of g because $df = \frac{1}{2}w_g - dH_g$, see (9.8).

The concept of Bäcklund transformations depends on the choice of affine coordinates. The following theorem clarifies this situation.

Theorem 6. *Let L be a Willmore surface in $\mathbb{H}P^1$. Choose non-zero $\beta \in (\mathbb{H}^2)^*, a \in \mathbb{H}^2$ such that $< \beta, a >= 0$. Then*

$$
d < \beta, *Aa >= 0 = d < \beta, *Qa > .
$$

If $g, h : M \to \mathbb{H} \subset \mathbb{H}P^1$ are immersions that satisfy

$$
dg = 2 < \beta, *Aa >, \quad dh = 2 < \beta, *Qa >,
$$

they are again Willmore surfaces, called forward *respectively* backward Bäcklund *transforms of L. The free choice of β implies that there is a whole S^4 of such pairs of Bäcklund transforms. (Different choices of a result in Moebius transforms $g \to g\lambda$, or $h \to h\lambda$, for a constant λ.)*

Proof. Choose $b \in \mathbb{H}^2$, $\alpha \in (\mathbb{H}^2)^*$ such that a, b and α, β are dual bases. Then

$$2 < \beta, *Aa >= \frac{1}{2}w, \qquad 2 < \beta, *Qa >= \frac{1}{2}w - dH,$$

see Proposition 12.

We can now proceed from g with another forward Bäcklund transform. To do so, we must integrate $\frac{1}{2}w_g = d(f + H_g)$. But, up to a translational constant, this yields

$$\tilde{f} := f + H_g. \tag{9.9}$$

We now observe

Lemma 10.

$$\begin{pmatrix} \tilde{f} \\ 1 \end{pmatrix} \in \ker A.$$

Proof. Note that $\ker A = \ker *A$. By Proposition 12 we have

$$4 * A \begin{pmatrix} \tilde{f} \\ 1 \end{pmatrix} = \begin{pmatrix} 1 & f \\ 0 & 1 \end{pmatrix} \begin{pmatrix} 0 & 0 \\ w\, dR + R * dR \end{pmatrix} \begin{pmatrix} 1 & -f \\ 0 & 1 \end{pmatrix} \begin{pmatrix} f + H_g \\ 1 \end{pmatrix}$$

$$= \begin{pmatrix} 1 & f \\ 0 & 1 \end{pmatrix} \begin{pmatrix} 0 & 0 \\ w\, dR + R * dR \end{pmatrix} \begin{pmatrix} H_g \\ 1 \end{pmatrix}$$

$$= \begin{pmatrix} 1 & f \\ 0 & 1 \end{pmatrix} \left(wH_g + \underbrace{dR + R * dR}_{=-dN_g + N_g * dN_g} \right)$$

$$= \begin{pmatrix} 1 & f \\ 0 & 1 \end{pmatrix} \begin{pmatrix} 0 \\ 2dgH_g - 2dgH_g \end{pmatrix} = 0.$$

Similarly the twofold *backward* Bäcklund transform $\hat{f}$ satisfies

$$\begin{pmatrix} \hat{f} \\ 1 \end{pmatrix} \mathbb{H} \supset \text{image}\, Q.$$

But this means that away from the zeros of A or Q the 2-step Bäcklund transforms of a Willmore surface L in $\mathbb{H}P^1$ can be described simply as $\tilde{L} = \ker A$ or $\hat{L} = \text{image}\, Q$. In particular there are no periods arising.

We obtain a chain of Bäcklund transforms

$$\cdots \to \hat{f} \to h \to f \to g \to \tilde{f} \to \cdots$$
$$\begin{array}{ccccccc} & & \| & & \| & & \| \\ \to & \hat{L} & \to & L & \to & \tilde{L} & \to \end{array}$$

Of course, the chain may break down if we arrive at non-immersed surfaces, or it may close up.

9.2 Two-Step Bäcklund Transforms

Let $L \subset H = M \times \mathbb{H}^2$ be a Willmore surface, and assume $A \not\equiv 0$ on each component of M. We want to describe directly the two-step Bäcklund transform $L \to \tilde{L}$, and compute its associated quantities (mean curvature sphere, Hopf fields).

We state a fact about singularities that will be proved in the appendix, see section 13.

Proposition 17. *Let L be a Willmore surface in $\mathbb{H}P^1$, and $A \not\equiv 0$ on each component of M. Then there exists a unique line bundle $\tilde{L} \subset H$ such that on an open dense subset of M we have:*

$$\tilde{L} = \ker A, \ \ and \ H = L \oplus \tilde{L}.$$

A similar assertion holds for image Q.

We shall assume that $\tilde{L}$ is immersed, and want to prove again that $\tilde{L}$ is Willmore.

Theorem 7. *For the 2-step Bäcklund transform $\tilde{L}$ of L we have*

$$\tilde{Q} = A. \tag{9.10}$$

Hence $\tilde{L}$ is again a Willmore surface.

Let $\tilde{S}, \tilde{\delta}, \tilde{Q}$, etc. denote the operators associated with $\tilde{L}$.

Lemma 11.

$$*\tilde{\delta} = -S\tilde{\delta}.$$

Proof. Since $A|_{\tilde{L}} = 0$ we interpret $A \in \Omega^1(\mathrm{Hom}(H/\tilde{L}, H))$. On a dense open subset of M then $A(X) : H/\tilde{L} \to H$ is injective for any $X \neq 0$.

For $\phi \in \Gamma(\tilde{L})$ we get

$$0 = d(*A)\phi = d(\underbrace{*A\phi}_{=0}) + *A \wedge d\phi = *A * d\phi + Ad\phi$$
$$= -AS * \tilde{\delta}\phi + A\tilde{\delta}\phi = -AS(*\tilde{\delta} + S\tilde{\delta})\phi.$$

The injectivity of A then proves the lemma.

Proof (of the theorem). Motivated by the lemma, we relate $\tilde{S}$ to $-S$ rather than to S. We put

$$\tilde{S} =: -S + B.$$

Then

$$4\tilde{Q} = \tilde{S}d\tilde{S} - *d\tilde{S}$$
$$= Bd\tilde{S} - (Sd\tilde{S} + *d\tilde{S})$$
$$= Bd\tilde{S} - (SdB + *dB) + (SdS + *dS)$$
$$= 4A + Bd\tilde{S} - (SdB + *dB).$$

The proof will be completed with the following lemma which shows that $\tilde{Q}$ – like A – has values in L, while the "B-terms" take values in $\tilde{L}$.

Lemma 12. *We have*

$$\text{image } B \subset \tilde{L}, \tag{9.11}$$
$$\text{image}(*dB + SdB) \subset \tilde{L}, \tag{9.12}$$
$$L \subset \ker B, \tag{9.13}$$
$$\text{image } \tilde{Q} \subset L. \tag{9.14}$$

Proof. Recall that $\tilde{L}$ is S-stable. It is of course also $\tilde{S}$-stable, and therefore

$$B\tilde{L} \subset \tilde{L}. \tag{9.15}$$

Now $\tilde{L}$ is immersive, and therefore image $\tilde{\delta} = H/\tilde{L}$. Thus (9.11) will follow if we can show $\tilde{\pi}Bd\phi = 0$ for $\phi \in \Gamma(\tilde{L})$. But, using Lemma 11,

$$\tilde{\pi}Bd\phi = \tilde{\pi}Sd\phi + \tilde{\pi}\tilde{S}d\phi = S\tilde{\pi}d\phi + \tilde{S}\tilde{\pi}d\phi = S\tilde{\delta}\phi + \tilde{S}\tilde{\delta}\phi$$
$$= -*\tilde{\delta}\phi + *\tilde{\delta}\phi = 0.$$

Next, for $\chi \in \Gamma(H)$ we have

$$\tilde{\pi}(*dB + SdB)\chi = \tilde{\pi}(*d(B\chi) + Sd(B\chi) - \underbrace{B*d\chi - SBd\chi}_{\tilde{L}-\text{valued}})$$
$$= (*\tilde{\delta} + S\tilde{\delta})B\chi$$
$$= 0. \qquad \text{(Lemma 11)}$$

This proves (9.12).

On the other hand, for $\psi \in \Gamma(L)$,

$$\tilde{\pi}(*dB - SdB)\psi = \tilde{\pi}\underbrace{(*dS - SdS)\psi}_{=-4Q\psi=0} + \tilde{\pi}(*d\tilde{S} - Sd\tilde{S})\psi$$
$$= \tilde{\pi}\underbrace{(*d\tilde{S} + \tilde{S}d\tilde{S})\psi}_{=4\tilde{A}\psi\in\Gamma(\tilde{L})} - \tilde{\pi}\underbrace{(Bd\tilde{S})\psi}_{\in\Gamma(\tilde{L})}$$
$$= 0.$$

Together with the previous equation we obtain $\tilde{\pi}dB|_L = 0$, and, for $\psi \in \Gamma(L)$,

$$\tilde{\delta}B\psi = \tilde{\pi}(d(B\psi)) = \tilde{\pi}((dB)\psi - Bd\psi) = \tilde{\pi}dB\psi = 0.$$

But $\tilde{L}$ is an immersion, and therefore $B\psi = 0$, proving (9.13).

Finally, for $\psi \in \Gamma(L)$,

$$
\begin{aligned}
4\tilde{Q}\psi &= \tilde{S}d\tilde{S}\psi - *d\tilde{S}\psi \\
&= \tilde{S}d\tilde{S}\psi - d\psi + \tilde{S} * d\psi - (-d\psi + \tilde{S} * d\psi + *d\tilde{S}\psi) \\
&= \tilde{S}(d\tilde{S}\psi + \tilde{S}d\psi + *d\psi) - *(*d\psi + \tilde{S}d\psi + d\tilde{S}\psi) \\
&= (\tilde{S} - *)(d(\tilde{S}\psi) + *d\psi) \\
&= -(\tilde{S} - *)(d(S\psi) - *d\psi) \quad \text{using (9.13)}.
\end{aligned}
$$

But $\pi(d(S\psi) - *d\psi) = (\delta S - *\delta)\psi = 0$. So $d(S\psi) - *d\psi \in \Gamma(L)$, and this is stable under $\tilde{S} = B - S$. Therefore $\tilde{Q}L \subset L$. Since $\tilde{Q}\tilde{L} = 0$, this proves (9.14).

Taking the two-step backward transform of $\tilde{L}$, we get image $\tilde{Q}$ = image A = L. Hence $\hat{\tilde{L}} = L$. We remark that the results of this section similarly apply to the backward two-step Bäcklund transformation $L \to \hat{L}$ = image Q. As a corollary of (9.10) and its analog $\hat{A} = Q$ we obtain

Theorem 8.

$$\hat{\tilde{L}} = L = \tilde{\hat{L}}.$$

10 Willmore Surfaces in S^3

Let $< .,>$ be an indefinite hermitian inner product on $\mathbb{H}^2$. To be specific, we choose

$$< v, w >:= \bar{v}_1 w_2 + \bar{v}_2 w_1.$$

Then the set of isotropic lines $< l, l >= 0$ defines an $S^3 \subset \mathbb{HP}^1$, while the complementary 4-discs are hyperbolic 4-spaces, see Example 4. We have

$$\begin{pmatrix} a & b \\ c & d \end{pmatrix}^* = \begin{pmatrix} \bar{d} & \bar{b} \\ \bar{c} & \bar{a} \end{pmatrix}, \tag{10.1}$$

and the same holds for matrix representations with respect to a basis (v, w) such that

$$< v, v >= 0 =< w, w >, \quad < v, w >= 1.$$

10.1 Surfaces in S^3

Let L be an isotropic line bundle with mean curvature sphere S. We look at the adjoint map $M \to \mathcal{Z}, p \mapsto S_p^*$ with respect to $< .,. >$. Clearly S^* stabilizes $L^\perp$, and $L = L^\perp$ implies

$$S^* L = S^* L^\perp = L^\perp = L.$$

Similarly,

$$(dS^*) L = (dS)^* L^\perp \subset L^\perp = L.$$

Moreover, if $Q^\dagger$ belongs to S^*, then

$$Q^\dagger = \frac{1}{4}(S^* dS^* - *dS^*)$$

$$= \frac{1}{4}(dSS - *dS)^*$$

$$= -\frac{1}{4}(SdS + *dS)^*$$

$$= -A^*.$$

Therefore $\ker Q^\dagger = (\mathrm{image}(Q^\dagger)^*)^\perp = (\mathrm{image}\,A)^\perp \supset L^\perp = L$.

We proceed to show that S and S^* coincide on L and H/L. By the uniqueness of the mean curvature sphere, see Theorem 2, it then follows that $S^* = S$.

Let $\psi \in \Gamma(L)$, and write

$$S\psi = \psi\lambda, \quad S^*\psi = \psi\mu$$

and

$$< \psi, \delta\psi > = \omega.$$

Note that $< \psi, \delta\psi >$ makes sense, because of $< \psi, L > = 0$. Differentiation of $\psi, \psi > = 0$ yields

$$\omega + \bar{\omega} = 0.$$

From $0 = < \psi, S\psi >$ we obtain

$$
\begin{aligned}
0 &= < \delta\psi, S\psi > + \underbrace{< \psi, (dS)\psi >}_{=0} + < \psi, S\delta\psi > \\
&= < \delta\psi, S\psi > + < S^*\psi, \delta\psi > \\
&= < \delta\psi, \psi > \lambda + \bar{\mu} < \psi, \delta\psi > \\
&= \bar{\omega}\lambda + \bar{\mu}\omega.
\end{aligned}
$$

Now we apply $*$ using

$$*\omega = < \psi, \delta S\psi > = \omega\lambda, \tag{10.2}$$

and obtain

$$0 = \bar{\lambda}\bar{\omega}\lambda + \bar{\mu}\omega\lambda = (\bar{\mu} - \bar{\lambda})\bar{\omega}\lambda.$$

We conclude $\lambda = \mu$, i.e. $S|_L = S^*|_L$.

Now assume $S|_{H/L} = I\rho$ and $S^*|_{H/L} = I\sigma$. Then

$$
\begin{aligned}
0 &= < \delta\psi, S\psi > + < S^*\psi, \delta\psi > \\
&= < S^*\delta\psi, \psi > + < \psi, S\delta\psi > \\
&= \bar{\sigma} < \delta\psi, \psi > + < \psi, \delta\psi > \rho \\
&= \bar{\sigma}\bar{\omega} + \omega\rho.
\end{aligned}
$$

But

$$
\begin{aligned}
*\omega &= < \psi, \delta S\psi > = < \psi, S\delta\psi > = \omega\rho \\
&= < S^*\psi, \delta\psi > = \bar{\lambda}\omega.
\end{aligned}
$$

Comparison with (10.2) shows $\rho = \lambda$, and we get

$$0 = \bar{\sigma}\overline{\lambda}\omega + \bar{\lambda}\omega\lambda = (\bar{\lambda} - \bar{\sigma})\omega\lambda.$$

It follows $\sigma = \lambda = \rho$, i.e. $S|_{H/L} = S^*|_{H/L}$. This completes the assumptions of Theorem 2, and $S^* = S$ by uniqueness.

Conversely, if $S^* = S$ and $S\psi = \psi\lambda$, then

$$\bar{\lambda} < \psi, \psi > = < S\psi, \psi > = < \psi, S\psi > = < \psi, \psi > \lambda = \lambda < \psi, \psi > .$$

Now $S^2 = -I$ implies $\lambda^2 = -1$, and therefore we get $< \psi, \psi > = 0$.

Proposition 18. *An immersed holomorphic curve L in $\mathbb{H}P^1$ is isotropic, i.e. a surface in S^3, if and only if $S = S^*$.*

10.2 Hyperbolic 2-Planes

In the half-space or Poincaré model of the hyperbolic space, geodesics are euclidean circles that orthogonally intersect the boundary. We consider the models of hyperbolic 4-space in $\mathbb{H}P^1$, and want to identify their totally geodesic hyperbolic 2-planes, i.e. those 2-spheres in $\mathbb{H}P^1$ that orthogonally intersect the separating isotropic S^3. Using the affine coordinates, from Example 4, we consider the reflexion $\mathbb{H} \to \mathbb{H}, x \mapsto -\bar{x}$ at $\operatorname{Im}\mathbb{H} = S^3$. This preserves either of the metrics given in the examples of section 3.2. In particular, it induces an isometry of the standard Riemannian metric of $\mathbb{H}P^1$ which fixes S^3. Given a 2-sphere $S \in \operatorname{End}(\mathbb{H}^2), S^2 = -I$, that intersects S^3 in a point l, we use affine coordinates, as in Example 4, with $l = v\mathbb{H}$ and w such that

$$< v, v > = < w, w > = 0, < v, w > = 1.$$

Then

$$S = \begin{pmatrix} N & -H \\ 0 & -R \end{pmatrix}$$

with $N^2 = R^2 = -1, NH = HR$, and $S' \subset \mathbb{H}$ is the locus of

$$Nx + xR = H.$$

If S' is invariant under the reflexion at S^3, then it also is the locus of $-N\bar{x} - \bar{x}R = H$ or

$$Rx + xN = \bar{H}.$$

According to section 3.4, the triple (H, N, R) is unique up to sign. This implies either

$$(H, N, R) = (\bar{H}, R, N) \quad \text{or} \quad (H, N, R) = (-\bar{H}, -R, -N).$$

By (10.1) either $S^* = S$, and the 2-sphere lies within the 3-sphere, or it intersects orthogonally, and $S^* = -S$. We summarize:

Proposition 19. *A 2-sphere $S \in \mathcal{Z}$ intersects the hyperbolic 4-spaces determined by an indefinite inner product in hyperbolic 2-planes if and only if $S^* = -S$.*

10.3 Willmore Surfaces in S^3 and Minimal Surfaces in Hyperbolic 4-Space

Let L be a connected Willmore surface in $S^3 \subset \mathbb{H}P^1$, where S^3 is the isotropic set of an indefinite hermitian form on $\mathbb{H}^2$. Then its mean curvature sphere satisfies

$$S^* = S.$$

Let us assume that $A \not\equiv 0$, and let $\tilde{L} = \ker A$ and $\hat{L} = \operatorname{image} Q$ be the 2-step Bäcklund transforms of L.

Lemma 13.

$$\hat{L} = \tilde{L}.$$

Proof. First we have

$$Q^* = \frac{1}{4}(SdS - *dS)^* = \frac{1}{4}(dSS - *dS)$$
$$= \frac{1}{4}(-SdS - *dS) = -A. \tag{10.3}$$

Now $\hat{L} = \operatorname{image} Q$ is S-stable, and $S^* = S$ and $S\phi = \phi\lambda$ imply $< \phi, \phi >= 0$. Therefore $< \hat{L}, \hat{L} >= 0$, and on a dense open subset of M

$$\hat{L} = \hat{L}^\perp = (\operatorname{image} Q)^\perp = \ker Q^* = \ker A = \tilde{L}.$$

Lemma 14.

$$\tilde{S} = -S$$

for the mean curvature sphere $\tilde{S}$ of $\tilde{L}$.

Proof. First $\tilde{L} = \hat{L}$ is obviously $(-S)$-stable. It is trivially invariant under A and Q and, therefore, under $d(-S) = 2(*A - *Q)$. Finally, the Q of $(-S)$ is

$$\frac{1}{4}((-S)d(-S) - *d(-S)) = A,$$

and this vanishes on $\tilde{L}$. The unique characterization of the mean curvature sphere by these three properties implies $\tilde{S} = -S$.

We now turn to the 1-step Bäcklund transform of L. If $dF = 2 * A$, then

$$d(F + F^*) = 2 * A + 2 * A^* \underset{(10.3)}{=} 2 * A - 2 * Q = -dS.$$

Because $S^* = S$, we can choose suitable initial conditions for F such that

$$F + F^* = -S. \tag{10.4}$$

We now use affine coordinates with $L = \begin{bmatrix} f \\ 1 \end{bmatrix}$. Then the lower left entry g of F is a Bäcklund transform of f, and (7.9) and (10.4) imply

$$g + \bar{g} = H.$$

We want to compute the mean curvature sphere S_g. From the properties of Bäcklund transforms we know

$$N_g = -R, \quad H_g = \tilde{f} - f, \tag{10.5}$$

see (9.3), (9.9). Likewise, $\tilde{N} = -R_g$. From Lemma 14 we obtain

$$\begin{pmatrix} 1 & f \\ 0 & 1 \end{pmatrix} \begin{pmatrix} -N & 0 \\ H & R \end{pmatrix} \begin{pmatrix} 1 & -f \\ 0 & 1 \end{pmatrix} = \begin{pmatrix} 1 & \tilde{f} \\ 0 & 1 \end{pmatrix} \begin{pmatrix} \tilde{N} & 0 \\ -\tilde{H} & -\tilde{R} \end{pmatrix} \begin{pmatrix} 1 & -\tilde{f} \\ 0 & 1 \end{pmatrix}$$

$$= \begin{pmatrix} 1 & f \\ 0 & 1 \end{pmatrix} \begin{pmatrix} 1 & H_g \\ 0 & 1 \end{pmatrix} \begin{pmatrix} \tilde{N} & 0 \\ -\tilde{H} & -\tilde{R} \end{pmatrix} \begin{pmatrix} 1 & -H_g \\ 0 & 1 \end{pmatrix} \begin{pmatrix} 1 & -f \\ 0 & 1 \end{pmatrix}$$

$$= \begin{pmatrix} 1 & f \\ 0 & 1 \end{pmatrix} \begin{pmatrix} \tilde{N} - H_g \tilde{H} & * \\ -\tilde{H} & * \end{pmatrix} \begin{pmatrix} 1 & -f \\ 0 & 1 \end{pmatrix}.$$

This implies $H = -\tilde{H}$ and $-N = \tilde{N} - H_g\tilde{H}$, whence

$$-R_g = \tilde{N} = -N + (f - \tilde{f})H.$$

In particular $f - \tilde{f} \in \operatorname{Im} \mathbb{H}$, since $H = 0$ on an open set would mean $w = 0$ on that set. It follows that

$$S_g = \begin{pmatrix} 1 & g \\ 0 & 1 \end{pmatrix} \begin{pmatrix} -R & 0 \\ f - \tilde{f} & -N + (f - \tilde{f})H \end{pmatrix} \begin{pmatrix} 1 & -g \\ 0 & 1 \end{pmatrix},$$

and, because $R = N$ and $H \in \mathbb{R}$ for $f : M \to \operatorname{Im} \mathbb{H} = \mathbb{R}^3$,

$$S_g^* = \begin{pmatrix} 1 & g - H \\ 0 & 1 \end{pmatrix} \begin{pmatrix} \overline{-N + (f - \tilde{f})H} & 0 \\ \overline{f - \tilde{f}} & -\bar{R} \end{pmatrix} \begin{pmatrix} 1 & H - g \\ 0 & 1 \end{pmatrix}$$

$$= \begin{pmatrix} 1 & g - H \\ 0 & 1 \end{pmatrix} \begin{pmatrix} N + (\tilde{f} - f)H & 0 \\ \tilde{f} - f & N \end{pmatrix} \begin{pmatrix} 1 & H - g \\ 0 & 1 \end{pmatrix}$$

$$= \begin{pmatrix} 1 & g \\ 0 & 1 \end{pmatrix} \begin{pmatrix} N & 0 \\ \tilde{f} - f & N + (\tilde{f} - f)H \end{pmatrix} \begin{pmatrix} 1 & -g \\ 0 & 1 \end{pmatrix}$$

$$= -S_g.$$

We have now shown that the mean curvature spheres of g intersect S^3 orthogonally, and therefore are hyperbolic planes. We know that, using affine coordinates and a Euclidean metric, the mean curvature spheres are tangent to g and have the same mean curvature vector as g. This property remains under conformal changes of the ambient metric. Therefore, in the hyperbolic metric, g has mean curvature 0, and hence is minimal. If $A \equiv 0$, then $w = 0$, and the "Bäcklund transform" is constant, which may be considered as a degenerate minimal surface. In general g will be singular in the (isolated) zeros of $dg = \frac{1}{2}w$, but minimal elsewhere.

We show the converse: Let L be an immersed holomorphic curve, minimal in hyperbolic 4-space, i.e. with $S^* = -S$. Then

$$A^* = \frac{1}{4}(SdS + *dS)^* = \frac{1}{4}(dSS - *dS) = -\frac{1}{4}(SdS + *dS) = -A,$$

and therefore also

$$(d * A)^* = -d * A.$$

From Proposition 15 we have

$$d * A = \frac{1}{4}\begin{pmatrix} f\,dw & -f\,dw\,f \\ dw & -dw\,f \end{pmatrix}.$$

Therefore

$$dw = -\overline{dw}, \qquad \overline{f\,dw} = dw\,f,$$

and hence

$$dw(f + \bar{f}) = 0.$$

But f is not in S^3, and therefore $dw=0$, i.e L is Willmore. Similarly, Proposition 12 yields

$$*A = \begin{pmatrix} * & * \\ w & * \end{pmatrix},$$

and $A^* = -A$ implies $w = -\bar{w}$. From $S^* = -S$ we know $\bar{H} = -H$, and the backward Bäcklund transform h with $dh = \frac{1}{2}w - dH$ and suitable initial conditions is in $\mathrm{Im}\,\mathbb{H} = \mathbb{R}^3$.

To summarize

Theorem 9 (Richter [11]). *Let $< .,. >$ be an indefinite hermitian product on $\mathbb{H}^2$. Then the isotropic lines form an $S^3 \subset \mathbb{H}P^1$, while the two complementary discs inherit complete hyperbolic metrics. Let L be a Willmore surface in $S^3 \subset \mathbb{H}P^1$. Then a suitable forward Bäcklund transform of L is hyperbolic minimal. Conversely, an immersed holomorphic curve that is hyperbolic minimal is Willmore, and a suitable backward Bäcklund transformation is a Willmore surface in S^3. (In both cases the Bäcklund transforms may have singularities.)*

11 Spherical Willmore Surfaces in $\mathbb{H}P^1$

In this chapter we sketch a proof of the following theorem of Ejiri [2] and Montiel [8], which generalizes an earlier result of Bryant [1] for Willmore spheres in S^3. See also Musso [9].

Theorem 10 (Ejiri [2], Montiel [8]). *A* Willmore *sphere in* $\mathbb{H}P^1$ *is a twistor projection of a holomorphic or anti-holomorphic curve in* $\mathbb{C}P^3$*, or, in suitable affine coordinates, corresponds to a minimal surface in* $\mathbb{R}^4$*.*

The material differs from what we have treated so far: The theorem is global, and therefore requires global methods of proof. These are imported from complex function theory.

11.1 Complex Line Bundles: Degree and Holomorphicity

Let E be a complex vector bundle. We keep the symbol $J \in \text{End}(H)$ for the endomorphism given by multiplication with the imaginary unit i.

We denote by $\bar{E}$ the bundle where J is replaced by $-J$. If $< .,. >$ is a hermitian metric on E, then

$$\bar{E} \to E^* = E^{-1}, \psi \to < \psi, . >$$

is an isomorphism of complex vector bundles. Also note that for complex *line* bundles E_1, E_2 the bundle $\text{Hom}(E_1, E_2)$ is again a complex line bundle.

There is a powerful integer invariant for complex line bundles E over a compact Riemann surface: the *degree*. It classifies these bundles up to isomorphism. Here are two equivalent definitions for the degree.

– Choose a hermitian metric $< .,. >$ and a compatible connection ∇ on E. Then $< R(X,Y)\psi, \psi >= 0$ for the curvature tensor R of ∇. Therefore $R(X,Y) = -\omega(X,Y)J$ with a real 2-form $\omega \in \Omega^2(M)$. Define

$$\deg(E) := \frac{1}{2\pi} \int_M \omega.$$

– Choose a section $\phi \in \Gamma(E)$ with isolated zeros. Then

$$\deg(E) := \operatorname{ord} \phi := \sum_{\phi(p)=0} \operatorname{ind}_p \phi.$$

The index of a zero p of ϕ is defined using a local non-vanishing section ψ and a holomorphic parameter z for M with $z(0) = p$. Then $\phi(z) = \psi(z)\lambda(z)$ for some complex function $\lambda : \mathbb{C} \subset U \to \mathbb{C}$ with isolated zero at 0, and

$$\operatorname{ind}_p \phi = \frac{1}{2\pi i} \int_\gamma \frac{d\lambda}{\lambda(z)},$$

where γ is a small circle around 0.

We state fundamental properties of the degree. We have

$$\deg(\bar{E}) = \deg E^{-1} = -\deg E,$$
$$\deg \operatorname{Hom}(E_1, E_2) = -\deg E_1 + \deg E_2.$$

More generally,

$$\deg(E_1 \otimes E_2) = \deg E_1 + \deg E_2.$$

Example 21. Let M be a compact Riemann surface of genus g, and E its tangent bundle, viewed as a complex line bundle. We compute its degree using the first definition. The curvature tensor of a surface with Riemannian metric $< \cdot, \cdot >$ is given by $R(X,Y) = K(< Y, \cdot > X - < X, \cdot > Y)$, where K is the Gaussian curvature. Let Z be a (local) unit vector field and $< \cdot, >$ compatible with J. Then

$$
\begin{aligned}
\omega(X,Y) &= \frac{1}{2} \operatorname{trace}_{\mathbb{R}} R(X,Y)J \\
&= \frac{K}{2}(< Y, JZ >< X, Z > - < X, JZ >< Y, Z > \\
&\quad - < Y, Z >< X, JZ > + < X, Z >< Y, JZ >) \\
&= K(< Y, JZ >< X, Z > - < X, JZ >< Y, Z >) \\
&= K \det \begin{pmatrix} < X, Z > & < X, JZ > \\ < Y, Z > & < Y, JZ >) \end{pmatrix} \\
&= K \, dA(X,Y).
\end{aligned}
$$

We integrate this using Gauss-Bonnet, and find $2\pi\chi(M) = 2\pi(2 - 2g) = 2\pi \deg(E)$. For the canonical bundle

$$K := E^{-1} = \operatorname{Hom}(TM, \mathbb{C}) = \{\omega \in \operatorname{Hom}_{\mathbb{R}}(TM, \mathbb{C}) \, | \, \omega(JX) = i\omega(X)\}$$

we therefore find

$$\deg(K) = 2g - 2.$$

Definition 11. *Let E be a complex vector bundle. A* holomorphic structure *for E is a complex linear map a map $\bar{\partial}$ from the sections of E into the E-valued complex anti-linear 1-forms $\bar{K}E$*

$$\bar{\partial} : \Gamma(E) \to \Gamma(\bar{K}E)$$

satisfying

$$\bar{\partial}(\lambda\psi) = (\bar{\partial}\psi)\lambda + \psi(\bar{\partial}\lambda).$$

Here $\bar{\partial}\lambda := \frac{1}{2}(d\lambda + i*d\lambda)$. (Local) sections $\psi \in \Gamma(E|_U)$ are called holomorphic, *if $\bar{\partial}\psi = 0$. We denote by $H^0(E|_U)$ the vector space of holomorphic sections over U.*

If E is a complex *line* bundle with holomorphic structure, and $\psi \in H^0(E)\backslash\{0\}$, then the zeros of ψ are isolated and of positive index because holomorphic maps preserve orientation. In particular, if M is compact and $\deg E < 0$, then any global holomorphic section in E vanishes identically.

In the proof of the Ejiri theorem we shall apply the concepts of degree and holomorphicity to several complex bundles obtained from quaternionic ones. We relate these concepts.

Definition 12. *If (L, J) is a complex quaternionic line bundle, then*

$$E_L := \{\psi \in L \mid J\psi = \psi i\}$$

is a complex line bundle. We define

$$\deg L := \deg E_L.$$

Lemma 15. *If L_1, L_2 are complex quaternionic line bundles, and $E_i := E_{L_i}$, then*

$$\mathrm{Hom}_+(L_1, L_2) \to \mathrm{Hom}_{\mathbb{C}}(E_1, E_2)$$
$$B \mapsto B|_{E_1}$$

is an isomorphism of complex vector bundles. In particular

$$\deg \mathrm{Hom}_+(L_1, L_2) = -\deg L_1 + \deg L_2.$$

The proof is straightforward. We now discuss one example in detail.

Example 22. We consider an immersed holomorphic curve

$$L \subset H = M \times \mathbb{H}^2$$

in $\mathbb{H}P^1$ with mean curvature sphere S. The bundle $K \mathrm{End}_-(H)$ is a complex vector bundle, the complex structure being given by post-composition with S. For $B \in \Gamma(K \mathrm{End}_-(H))$ we define

$$(\bar{\partial}_X B)(Y)\psi = \bar{\partial}_X(B(Y)\psi) - B(\bar{\partial}_X Y)\psi - B(Y)\partial_X\psi,$$

where

$$\bar{\partial}_X Y := \frac{1}{2}([X, Y] + J[JX, Y]),$$

$$\bar{\partial}\psi = \frac{1}{2}(d + S * d)\psi, \quad \partial\psi = \frac{1}{2}(d - S * d)\psi \text{ for } \psi \in \Gamma(H).$$

Direct computation shows that this is in fact a holomorphic structure, namely that induced on

$$K \operatorname{End}_-(H) = K \operatorname{Hom}_+(\bar{H}, H) = K \operatorname{Hom}_{\mathbb{C}}(\bar{H}, H)$$

by $\bar{\partial}$ on TM, and the above (quaternionic) holomorphic structures $\bar{\partial}$ on H and ∂ on $\bar{H}$.

Lemma 16.

$$(d * A)(X, JX) = -2(\bar{\partial}_X A)(X).$$

Proof. Let X be a local holomorphic vector field, i.e. $[X, JX] = 0$, see Remark 12, and $\psi \in \Gamma(H)$. Then

$$(d * A)(X, JX)\psi = (-X \cdot A(X) - (JX) \cdot SA(X) - A(\underbrace{[X, JX]}_{=0}))\psi$$

$$= -(d(\underbrace{A(X)\psi}_{=:\phi}) + *d(SA(X)\psi))(X)$$

$$+ A(X)d\psi(X) + SA(X) * d\psi(X)$$

$$= -(d\phi + *d(S\phi))(X) + A(X)(d\psi - S * d\psi)(X).$$

Now

$$d\phi + *d(S\phi) = (\partial + \bar{\partial} + A + Q)\phi + *(\partial + \bar{\partial} + A + Q)S\phi$$
$$= (\partial + \bar{\partial} + A + Q)\phi + (S\partial - S\bar{\partial} + SA - SQ)S\phi$$
$$= (\partial + \bar{\partial} + A + Q)\phi + (-\partial + \bar{\partial} + A - Q)\phi$$
$$= 2(\bar{\partial} + A)\phi$$
$$= 2\bar{\partial}(A(X)\phi) + 2AA(X)\phi.$$

Similarly

$$d\psi - S * d\psi = (\partial + \bar{\partial} + A + Q)\psi - S * (\partial + \bar{\partial} + A + Q)\psi$$
$$= (\partial + \bar{\partial} + A + Q)\psi - S(S\partial - S\bar{\partial} + SA - SQ)\psi$$
$$= (\partial + \bar{\partial} + A + Q)\psi - (-\partial + \bar{\partial} - A + Q)\psi$$
$$= 2(\partial + A)\psi.$$

Therefore

$$(d * A)(X, JX)\psi = -2\bar{\partial}_X(A(X)\psi) - 2A(X)^2\psi + 2A(X)\partial_X\psi + 2A(X)^2\psi$$
$$= -2(\bar{\partial}_X(A(X)\psi) - A(X)\partial_X\psi)$$
$$= -2(\bar{\partial}_X A)(X)\psi.$$

Now assume that L is Willmore, and therefore $d * A = 0$. This implies $\bar{\partial}A = 0$, and A is holomorphic:

$$A \in H^0(K \operatorname{End}_-(H)) = H^0(K \operatorname{Hom}_+(\bar{H}, H)).$$

As a consequence, see Lemma 23, either $A \equiv 0$, or the zeros of A are isolated, and there exists a line bundle $\tilde{L} \subset H$ such that $\tilde{L} = \ker A$ away from the zeros of A. For local $\psi \in \Gamma(\tilde{L})$ and holomorphic $Y \in H^0(TM)$ we have

$$\underbrace{\bar{\partial}A}_{=0}(Y)\psi = \bar{\partial}\underbrace{(A(Y)\psi)}_{=0} - A(Y)\partial\psi.$$

Therefore $\tilde{L}$ is invariant under ∂, like L is invariant under $\bar{\partial}$, see Remark 6. As above, we get a holomorphic structure on the complex *line* bundle $K \operatorname{Hom}_+(\bar{H}/\tilde{L}, L)$ and A defines a holomorphic section of this bundle:

$$A \in H^0(K \operatorname{Hom}_+(\bar{H}/\tilde{L}, L)).$$

11.2 Spherical Willmore Surfaces

We turn to the

Proof (of Theorem 10). If $A \equiv 0$ or $Q \equiv 0$, then L is a twistor projection by Theorem 5.

Otherwise we have the line bundle $\tilde{L}$, and similarly a line bundle $\hat{L}$ that coincides with the image of Q almost everywhere.

Proposition 20. *We have the following holomorphic sections of complex holomorphic line bundles:*

$$A \in H^0(K \operatorname{Hom}_+(\bar{H}/\tilde{L}, L)), \quad Q \in H^0(K \operatorname{Hom}_+(H/L, \bar{\tilde{L}})),$$
$$\delta_L \in H^0(K \operatorname{Hom}_+(L, H/L)), \quad AQ \in H^0(K^2 \operatorname{Hom}_+(H/L, L))$$
$$\text{and if } AQ = 0 \text{ then} \qquad \delta_{\tilde{L}} \in H^0(K \operatorname{Hom}_+(\bar{\tilde{L}}, \bar{H}/\tilde{L}))$$

We proved the statement about A. We give the (similar) proofs of the others in the appendix.

The degree formula then yields

$$\operatorname{ord}\delta_L = \deg K - \deg L + \deg H/L$$
$$\operatorname{ord}(AQ) = 2\deg K - \deg H/L + \deg L$$
$$= 3\deg K - \operatorname{ord}\delta_L$$
$$= 6(g-1) - \operatorname{ord}\delta_L.$$

For $M = S^2$, i.e. $g = 0$, we get $\operatorname{ord}(AQ) < 0$, whence $AQ = 0$. Then $\tilde{L} = \hat{L}$, and

$$\operatorname{ord}A = \deg K + \deg H/\tilde{L} + \deg L$$
$$\operatorname{ord}Q = \deg K - \deg H/L - \deg \tilde{L}$$
$$\operatorname{ord}\delta_{\tilde{L}} = \deg K + \deg \tilde{L} - \deg H/\tilde{L}.$$

Addition yields

$$\operatorname{ord}\delta_{\tilde{L}} + \operatorname{ord}Q + \operatorname{ord}A = 3\deg K - \deg H/L + \deg L$$
$$= 4\deg K - \operatorname{ord}\delta_L = -8 - \operatorname{ord}\delta_L.$$

It follows that $\operatorname{ord}\delta_{\tilde{L}} < 0$, i.e. $\delta_{\tilde{L}} = 0$, and $\tilde{L}$ is d-stable, hence constant in $H = M \times \mathbb{H}^2$. From $AS = -SA = 0$ we conclude $S\tilde{L} = \tilde{L}$. Therefore all mean curvature spheres of L pass through the fixed point $\tilde{L}$. Choosing affine coordinates with $\tilde{L} = \infty$, all mean curvature spheres are affine planes, and L corresponds to a minimal surface in $\mathbb{R}^4$.

12 Darboux tranforms

Bäcklund transforms provided a mean to construct new Willmore surfaces out of a given one by solving linear equations. Darboux transforms provide another method for such construction, based on the solution of a Riccati equation. For isothermic surfaces it is described in [6].

After an introductory remark on Riccati equations, we first describe Darboux transforms for a special case of [6], namely for constant mean curvature surfaces in $\mathbb{R}^3$, because it displays a striking similarity with the Willmore case treated thereafter.

As with the Bäcklund transforms the theory – in the Willmore case – is again local. We only have a local existence of a solution to the Riccati initial value problem, and moreover require this solution to be invertible in the algebra $\text{End}(\mathbb{H}^2)$.

12.1 Riccati equations

We consider Riccati type partial differential equations in an algebra, which below will be $\mathbb{H}$ or $\text{End}(\mathbb{H}^2)$.

Lemma 17. *Let $\mathcal{A}$ be an associative unitary algebra over the reals, and M a manifold. Let $\alpha, \beta \in \Omega^1(M, \mathcal{A})$ with*

$$d\alpha = 0 = d\beta,$$
$$\alpha \wedge \beta = 0 = \beta \wedge \alpha.$$

Then for any $\rho \in \mathbb{R}\backslash\{0\}, p_0 \in M$ and $T_0 \in \mathcal{A}$ the Riccati initial value problem

$$dT = \rho T \alpha T - \beta, \quad T(p_0) = T_0 \tag{12.1}$$

has a unique solution T on a connected neighborhood of p_0.
   Moreover, if $S : M \to \mathcal{A}$ with

$$S^2 = -1, \quad S\alpha + \alpha S = 0, \quad dS = \alpha - \beta,$$

and

$$(T - S)^2(p_0) = \rho^{-1} - 1,$$

then

$$(T - S)^2 = \rho^{-1} - 1 \qquad (12.2)$$

everywhere, and $T^2 S = S^2 T$.

Proof. The integrability condition for (12.1)

$$\begin{aligned}
0 &= \rho dT \wedge \alpha T + \rho T d\alpha T - \rho T \alpha \wedge dT - d\beta \\
&= \rho(\rho T \alpha T - \beta) \wedge \alpha T + \rho T d\alpha T - \rho T \alpha \wedge (\rho T \alpha T - \beta) - d\beta \\
&= -\rho \beta \wedge \alpha T + \rho T d\alpha T + \rho T \alpha \wedge \beta - d\beta
\end{aligned}$$

is obviously satisfied. Now, if T is a solution, and S as above, then we define

$$X := \rho(T - S)^2 + \rho - 1.$$

Then X satisfies a linear first order differential equation

$$\begin{aligned}
\rho^{-1} dX &= (dT - dS)(T - S) + (T - S)(dT - dS) \\
&= (\rho T \alpha T - \alpha)(T - S) + (T - S)(\rho T \alpha T - \alpha) \\
&= T\alpha(\rho T^2 - \rho T S - 1) + (\rho T^2 - \rho S T - 1)\alpha T + \underbrace{\alpha S + S\alpha}_{=0} \\
&= T\alpha X + X\alpha T.
\end{aligned}$$

Hence $X(p_0) = 0$ implies $X = 0$. The last equation of the lemma follows from

$$T^2 S - ST^2 = (T - S)^2 S - S(T - S)^2$$

together with (12.2).

12.2 Constant mean curvature surfaces in $\mathbb{R}^3$

Let $f : M \to \mathrm{Im}\,\mathbb{H}$ be a conformal immersion:

$$*df = Ndf = -dfN, \quad N^2 = -1.$$

dN is a "'tangential"' 1-form: it anticommutes with N. We decompose it into the K- and $\bar{K}$-part with respect to the complex structure given by left multiplication by N to obtain

$$dN = Hdf + \omega. \qquad (12.3)$$

Since that is also the decomposition of the shape operator into "trace" and traceless part, the function $H : M \to \mathbb{R}$ is the mean curvature, and $*\omega = -N\omega$. Then

$$\omega = \frac{1}{2}(dN + N * dN).$$

Note that (12.3) resembles the formula $dS = 2 * Q - 2 * A$.

Now $-dN$ is the shape operator of f, and (12.3) gives its decomposition into the "trace" and the traceless part. Therefore $H : M \to \mathbb{R}$ is the *mean curvature*. Differentiating (12.3) we get

$$0 = dH \wedge df + d\omega.$$

Hence H is constant if and only if $d\omega = 0$, resembling $d * Q = 0$.

We see that the theory of constant mean curvature (=cmc) surfaces in $\mathbb{R}^3$ parallels that of Willmore surfaces in $\mathbb{HP}^1$.

We now assume H to be constant $\neq 0$. Then

$$g := f - \frac{1}{H}N$$

satisfies

$$dg = df - \frac{1}{H}dN = -\frac{\omega}{H},$$
$$dN = H(df - dg),$$
$$*dg = -N\,dg = dg\,N,$$

and

$$df \wedge dg = 0 = dg \wedge df$$

by type. The map g is an immersion of constant mean curvature H away from the umbilics of f, i.e. away from $\omega = 0$. It is called *the parallel constant mean curvature surface* of f.

For simplicity we restrict ourselves to the case

$$H = -1.$$

(The general case can be reduced to this using the homothety $f \to \mu f, H \to \frac{H}{\mu}, g \to \mu g$ with $\mu = -H$.)

We put $\mathcal{A} := \text{End}(\mathbb{H}), \alpha = dg, \beta = df$. These match the assumptions of lemma 17. Therefore, for any $\rho \neq 0, p_0 \in M$ and $T_0 \in \text{Im}\,\mathbb{H} \backslash \{0\}$ the initial value problem

$$dT = \rho T\,dg\,T - df, \quad T(p_0) = T_0$$

(locally) has unique solution T, which we assume to have no zeros. T stays in $\text{Im}\,\mathbb{H}$, because $\bar{T}$ satisfies the same equation up to a minus sign.

We put

$$f^\sharp = f + T.$$

Then

$$*df^\sharp = *(df + dT) = \rho T * dgT = -\rho TNdgT = -TNT^{-1}\rho TdgT$$
$$= -TNT^{-1}(df + dT) = -TNT^{-1}df^\sharp.$$

This shows that $f^\sharp$ is conformal with $N^\sharp := N_{f^\sharp} = -TNT^{-1}$. Moreover, $f^\sharp$ is an immersion if and only if g is an immersion. if g is immersive.

Under what conditions does $f^\sharp$ again have constant mean curvature? We compute $H^\sharp := H_{f^\sharp}$, using

$$T^2 = -|T|^2, \quad TN + NT = TN + \overline{TN} = -2 < T, N >,$$

and

$$dN \wedge df = Hdf \wedge df - Hdg \wedge df = Hdf \wedge df.$$

We find

$$
\begin{aligned}
H^\sharp df^\sharp \wedge df^\sharp &= dN^\sharp \wedge df^\sharp \\
&= -d(TNT^{-1}) \wedge df^\sharp \\
&= -(dTNT^{-1} + TdNT^{-1} - TNT^{-1}dTT^{-1}) \wedge \rho TdgT \\
&= -(dTN + TdN - TNT^{-1}dT) \wedge \rho dgT \\
&= (-(\rho TdgT - df)N - Tdg + TNT^{-1}(\rho TdgT - df)) \wedge \rho dgT \\
&= -\rho(Tdg(TN + NT) + \rho^{-1}Tdg)) \wedge \rho dgT \\
&= -\frac{2 < T, N > -\rho^{-1}}{|T|^2} \, \rho TdgT \wedge \rho TdgT \\
&= -\frac{2 < T, N > -\rho^{-1}}{|T|^2} \, df^\sharp \wedge df^\sharp.
\end{aligned}
$$

Hence we proved

Lemma 18.

$$H^\sharp = -\frac{2 < T, N > -\rho^{-1}}{|T|^2}.$$

Next we show

Lemma 19. *If $H^\sharp$ is constant, then $H^\sharp = -1$.*

Proof. We differentiate $0 = H^\sharp|T|^2 + 2 < T, N > -\rho^{-1}$:

$$0 = H^\sharp < dT, T > + < dT, N > + < T, dN >$$
$$= H^\sharp(< \rho TdgT, T > - < df, T >) + < \rho TdgT, N > - < T, df > + < T, dg >$$
$$= H^\sharp(-|T|^2\rho < T, dg >) - H^\sharp < df, T > + < \rho TdgT, N >$$
$$\quad - < T, df > + < T, dg >$$
$$= -(H^\sharp|T|^2\rho - 1) < T, dg > -(H^\sharp + 1) < T, df > + < \rho TdgT, N >$$
$$= 2\rho < T, N >< T, dg > -(H^\sharp + 1) < T, df > + < \rho TdgT, N >$$
$$= -(H^\sharp + 1) < T, df >$$
$$\quad + \frac{\rho}{2}((TN + NT)(Tdg + dgT) - (TdgTN + NTdgT))$$
$$= -(H^\sharp + 1) < T, df >$$
$$\quad + \frac{\rho}{2}(Tdg(TN + NT) + (TN + NT)dgT - (TdgTN + NTdgT))$$
$$= -(H^\sharp + 1) < T, df >$$
$$\quad + \frac{\rho}{2}(TdgTN + \underbrace{TdgNT + TNdgT}_{=0} + NTdgT - (TdgTN + NTdgT))$$
$$= -(H^\sharp + 1) < T, df > .$$

If $H^\sharp = -1$, we are done. Otherwise $< T, df >= 0$, i.e. $T = \mu N$, and

$$dT = d\mu N + \mu dN = d\mu N - \mu df + \mu\omega$$
$$dT = \rho TdgT - df = \rho\mu^2 NdgN - df = \rho\mu^2\omega - df$$

Now df and ω are tangential, and comparison of the above two equations gives

$$d\mu = 0,$$
$$(1 - \mu)df = (-\mu + \rho\mu^2)\omega,$$

$d\mu = 0$ and therefore

$$\mu = 1.$$

But then $f^\sharp = g$ is the parallel constant mean curvature surface of f which has $H^\sharp = -1$.

As a consequence of the preceeding two results we obtain

Lemma 20. $H^\sharp$ *is constant, if and only if*

$$(T - N)^2 = \rho^{-1} - 1. \tag{12.4}$$

Proof. We know that $H^\sharp$ ist constant, if and only if it equals -1, and this is equivalent with

$$|T|^2 - 2 <T, N> +\rho^{-1} = 0.$$

But

$$(T - N)^2 = -|T - N|^2 = -(|T|^2 - 2 <T, N> +1)$$
$$= -(|T|^2 - 2 <T, N> +\rho^{-1}) + \rho^{-1} - 1.$$

Now recall from lemma 17 that (12.4) holds everywhere, if it holds in a single point. Therefore $|T - S|^2 = 1 - \rho^{-1}$, and T is bounded with no zeros. Hence it can be globally defined if M is simpli connected. This leads us to the following

Definition 13. *Let* $f : M \to \operatorname{Im} \mathbb{H}$ *be a conformal immersion with constant mean curvature* $H = -1$, *and immersed parallel cmc surface* $g = f + N$. *Let*

$$\rho \in \mathbb{R}\backslash\{0\}, \quad T_0 \in \operatorname{Im} \mathbb{H}\backslash\{0\}, \quad p_0 \in M,$$

and assume

$$(T_0 - N(p_0))^2 = \rho^{-1} - 1. \tag{12.5}$$

Let T *be the unique solution of the Riccati initial value problem*

$$dT = \rho T dg T - df, \quad T(p_0) = T_0.$$

Then

$$f^\sharp := f + T$$

is called a Darboux transform *of* f.

Remark 11.   1. If H is constant $\neq 0, -1$, then (12.5) in the above definition should be replaced by

$$(HT_0 + N(p_0))^2 = \frac{H^2}{\rho} - 1.$$

It turns out that $f + T$ has again constant mean curvature H.
   2. From (12.5) we see that for a given $\rho \neq 0$ there is an S^2 of initial T_0 . Hence there is a 3-parameter family of Darboux transforms.

We summarize the previous results:

Theorem 11. *The Darboux transforms of surfaces with constant mean curvature* H *in* $\mathbb{R}^3$ *have constant mean curvature* H.

12.3 Darboux transforms of Willmore surfaces

Let $L \subset H = M \times \mathbb{H}^2$ be a Willmore surface in $\mathbb{H}P^1$ with mean curvature sphere S, and $dS = 2(*A - *Q)$. Since $d*A = d*Q = 0$ we can define $\text{End}(\mathbb{H}^2)$-valued maps F, G locally on M by

$$dF = 2 * A, \quad G = F + S.$$

Then

$$dG = 2 * Q,$$
$$dS = dG - dF,$$
$$dG \wedge dF = 0 = dF \wedge dG.$$

Hence the integrability conditions for the Riccati equation in $\mathcal{A} = \text{End}(\mathbb{H}^2)$, with $\alpha = dG, \beta = dF$ are satisfied.

As in the cmc case we find for any

$$\rho \in \mathbb{R} \backslash \{0\}, \quad T_0 \in GL(2, \mathbb{H}), \quad p_0 \in M$$

a unique (local) solution T of the Riccati initial value problem

$$dT = \rho T dG \, T - dF, \quad T(p_0) = T_0,$$

which we may assume to be invertible. As above let us assume that

$$(T_0 - S(p_0))^2 = (\rho^{-1} - 1)I.$$

Then

$$(T - S)^2 = (\rho^{-1} - 1)I$$

everywhere by lemma 17, and we call

$$L^\sharp := T^{-1} L$$

a *Darboux transform* of L.

Our aim now is to show that $L^\sharp$ is again Willmore. We start by computing its mean curvature sphere.

Lemma 21. *The mean curvature sphere of $L^\sharp$ is given by*

$$S^\sharp := T^{-1} S T = T S T^{-1},$$

and the corresponding Hopf fields are

$$2 * A^\sharp := \rho^{-1} T^{-1} dF \, T^{-1}, \quad 2 * Q^\sharp := \rho T dG \, T.$$

Proof. First note that the derivative $\delta^\sharp \in \Omega^1(\mathrm{Hom}(L^\sharp, H/L^\sharp))$ of $L^\sharp$ is given by

$$\delta^\sharp = T^{-1}\delta T.$$

Therefore $L^\sharp$ is immersed and

$$*\delta^\sharp = T^{-1} * \delta T = T^{-1}S\delta T = T^{-1}STT^{-1}\delta T = S^\sharp\delta^\sharp.$$

A similar computation yields $*\delta^\sharp = \delta^\sharp S^\sharp$. Due to the definition of $S^\sharp$ and $L^\sharp$ we obviously have

$$S^\sharp L^\sharp = L^\sharp.$$

Moreover,

$$T^{-1}ST = T^{-1}ST^2T^{-1} = T^{-1}T^2ST^{-1} = TST^{-1}.$$

Now

$$
\begin{aligned}
dS^\sharp &= d(TST^{-1})\\
&= dT\,ST^{-1} + T\,dS\,T^{-1} - TST^{-1}dT\,T^{-1}\\
&= (\rho T\,dGT - dF)ST^{-1} + T\,dS\,T^{-1} - TST^{-1}(\rho T\,dGT - dF)T^{-1}\\
&= T((\rho\,dGT - T^{-1}dF)S + dS - S(\rho\,dGT - T^{-1}dF))T^{-1}\\
&= T(\rho\,dG(TS + ST + \rho^{-1}I) + (T^{-1}S + ST^{-1} - 1)dF)T^{-1}\\
&= T(\rho\,dG\,T^2 - \rho^{-1}T^{-2}dF)T^{-1}\\
&= T\rho\,dG\,T - \rho^{-1}T^{-1}dF\,T^{-1}\\
&= 2(*Q^\sharp - *A^\sharp),
\end{aligned}
$$

which is the decomposition of $dS^\sharp$ into type:

$$*T\,dG\,T = T*(2*Q)T = -TS(2*Q)T = -TST^{-1}T\,dG\,T = -S^\sharp T\,dGT,$$

and similarly for F.

Finally,

$$Q^\sharp|_{L^\sharp} = 0,$$

and $A^\sharp \mathbb{H}^2 \subset L^\sharp$, whence

$$dS^\sharp L^\sharp \subset L^\sharp.$$

This proves that $S^\sharp$ is the mean curvature sphere of $L^\sharp$.

Theorem 12. *The Darboux transforms of an immersed Willmore surface in* $\mathbb{HP}^1$ *are again Willmore surfaces.*

Proof.

$$
\begin{aligned}
-2\rho^{-1}d * Q^{\sharp} &= d(TdGT) = dT \wedge dGT - TdG \wedge dT \\
&= (\rho TdGT - dF) \wedge dGT - TdG \wedge (\rho TdGT - dF) \\
&= \rho(TdGT \wedge dG - TdG \wedge TdGT) = 0.
\end{aligned}
$$

13 Appendix

13.1 The bundle $\tilde{L}$

Lemma 22. *If L is is an immersed holomorphic curve in $\mathbb{HP}^1$, then*

$$A|_L = 0 \iff A = 0.$$

Proof. Let $A|_L = 0$. Since $Q|_L = 0$, we find for $\psi \in \Gamma(L)$

$$0 = d(*Q\psi) = (d*Q)\psi - *Q \wedge d\psi = (d*Q)\psi \underset{(5.2)}{=} (d*A)\psi.$$

Note that $*Q \wedge d\psi = *Q \wedge \delta\psi = 0$ by type, where $Q : H \to H/L$. Since $A|_L = 0$, we now obtain similarly

$$0 = d(*A\psi) = \underbrace{(d*A)\psi}_{=0} - *A \wedge d\psi = -*A \wedge \delta\psi = -*A*\delta\psi - A\delta\psi = -2A\delta\psi.$$

But L is an immersion. Therefore $A|_L = 0 = A\delta$ implies $A = 0$. The converse is obvious.

Lemma 23. *Given a holomorphic section $T \in H^0(\mathrm{Hom}(V, W))$, where V, W are holomorphic complex vector bundles, there exist holomorphic subbundles*

$$V_0 \subset V, \hat{W} \subset W$$

such that $V_0 = \ker T$ and $\hat{W} = \mathrm{image}\, T$ away from a discrete subset.

Proof. Let $r := \max\{\mathrm{rank}\, T_p \,|\, p \in M\}$ and $G := \{p \,|\, \mathrm{rank}\, T_p = r\}$. This is an open subset of M. Let p_0 be a boundary point of G, an let $\psi_1, \ldots, \psi_n$ be holomorphic sections of V on a neighborhood U of p_0. By a change of indices we may assume that $T\psi_1 \wedge \ldots \wedge T\psi_r \not\equiv 0$. But this is a holomorphic section of the holomorphic bundle $\Lambda^r W|_U$, and hence has isolated zeros, because $\dim_{\mathbb{C}} M = 1$. We assume that p_0 is its only zero within U. Moreover, there exist $k \in \mathbb{N}$, a holomorphic coordinate z centered at p_0, and a holomorphic section $\sigma \in H^0(\Lambda^r W|_U)$ such that

$$T\psi_1 \wedge \ldots \wedge T\psi_r = z^k \sigma.$$

Off p_0 the section σ is decomposable, and since the Grassmannian $G_r(W)$ is closed in $\Lambda^r(W)$, it defines a section of $G_r(W)$, i.e. an r-dimensional subbundle of $W|_U$ extending image $T|_{U \setminus p_0}$. The statement about the kernel follows easily using the fact that $\ker T$ is the annihilator of image $T^* : W^* \to V^*$.

Proposition 21. *Let L be a (connected) Willmore surface in $\mathbb{H}P^1$, and $A \not\equiv 0$. Then there exists a unique line bundle $\tilde{L} \subset H$ such that on an open dense subset of M we have:*

$$\tilde{L} = \ker A \text{ and } H = L \oplus \tilde{L}.$$

Proof. $A \in \Gamma(K \operatorname{End}_-(H))$ is a holomorphic section by Example 22. By Lemma 23 there exists a line bundle $\tilde{L}$ such that $\tilde{L} = \ker A$ off a discrete set. Assume now that $H_p \neq L_p \oplus \tilde{L}_p$ for all p in an open non-empty set $U \subset M$. Then $L = \tilde{L}$, and $A|_L = 0$ on U. But then $A|_U = 0$ by Lemma 22. This is a contradiction, because the zeros of A are isolated.

13.2 Holomorphicity and the Ejiri/Montiel theorem

In this section L denotes an immersed holomorphic curve in $\mathbb{H}P^1$.

Remark 12 (Holomorphic Vector Fields). The tangent bundle of a Riemann surface viewed as complex line bundle carries a holomorphic structure:

$$\bar{\partial}_X Y = \frac{1}{2}([X, Y] + J[JX, Y]).$$

Note that this is tensorial in X. The vanishing of the Nijenhuis tensor implies $\bar{\partial} J = 0$. A vector field Y is called holomorphic if $\bar{\partial} Y = 0$. This is equivalent with $\bar{\partial}_Y Y = 0 = \bar{\partial}_{JY} Y$, but either of these conditions simply says

$$[Y, JY] = 0.$$

Any constant vector field in $\mathbb{C}$ is therefore holomorphic, and a given tangent vector to a Riemann surface can always be extended to a holomorphic vector field.

Proposition 22. *Let L be a Willmore surface in $\mathbb{H}P^1$. We have the following holomorphic sections of complex holomorphic line bundles:*

$$A \in H^0(K \operatorname{Hom}_+(\bar{H}/\tilde{L}, L)), \quad Q \in H^0(K \operatorname{Hom}_+(H/L, \bar{\tilde{L}})),$$

$$\delta_L \in H^0(K \operatorname{Hom}_+(L, H/L)), \quad AQ \in H^0(K^2 \operatorname{Hom}_+(H/L, L)),$$

$$\text{and if } AQ = 0 \text{ then} \quad \delta_{\tilde{L}} \in H^0(K \operatorname{Hom}_+(\tilde{L}, \bar{H}/\tilde{L})).$$

For the proof we need

Lemma 24. *The curvature tensor of the connection $\partial + \bar{\partial}$ on H is given by*

$$R^{\partial + \bar{\partial}} = -(A \wedge A + Q \wedge Q), \tag{13.1}$$

and for a holomorphic vector field Z we have

$$R^{\partial + \bar{\partial}}(Z, JZ) = 2S(\bar{\partial}_Z \partial_Z - \partial_Z \bar{\partial}_Z). \tag{13.2}$$

Proof. In general, if ∇ and $\tilde{\nabla} = \nabla + \omega$ are two connections, then

$$R^{\tilde{\nabla}} = R^{\nabla} + d^{\nabla}\omega + \omega \wedge \omega.$$

We apply this to $\tilde{\nabla} = \partial + \bar{\partial} = d - (A + Q)$ and use Lemma 4:

$$\begin{aligned}
R^{\partial + \bar{\partial}} &= R^d - d(A + Q) + (A + Q) \wedge (A + Q) \\
&= -2(A \wedge A + Q \wedge Q) + (A \wedge A + Q \wedge Q) \\
&= -(A \wedge A + Q \wedge Q).
\end{aligned}$$

Equation (13.2) follows from

$$\begin{aligned}
R^{\partial + \bar{\partial}}(Z, JZ) &= (\partial_Z + \bar{\partial}_Z)(\partial_{JZ} + \bar{\partial}_{JZ}) - (\partial_{JZ} + \bar{\partial}_{JZ})(\partial_Z + \bar{\partial}_Z) \\
&= S(\partial_Z + \bar{\partial}_Z)(\partial_Z - \bar{\partial}_Z) - S(\partial_Z - \bar{\partial}_Z)(\partial_Z + \bar{\partial}_Z) \\
&= 2S(-\partial_Z \bar{\partial}_Z + \bar{\partial}_Z \partial_Z),
\end{aligned}$$

because $\bar{\partial}_Z^2 = 0 = \partial_Z^2$.

Proof (Proof of the proposition).
   The holomorphicity of A was shown in example 22, and that of Q can be shown in complete analogy.
   (H, S) is a holomorphic complex quaternionic vector bundle, and L is a holomorphic subbundle, see Remark 6. Therefore L and H/L are holomorphic complex quaternionic line bundles, and the complex line bundle $K \operatorname{Hom}_+(L, E/L)$ inherits a holomorphic structure. Then, for local holomorphic sections ψ in L and Z in TM,

$$\begin{aligned}
(\bar{\partial}_Z \delta_L)(Z)\psi &= \bar{\partial}_Z(\delta_L(Z)\psi) - \delta_L(\bar{\partial}_Z Z)\psi - \delta_L(Z)(\bar{\partial}_Z \psi) \\
&= \bar{\partial}_Z(\delta_L(Z)\psi) = \bar{\partial}_Z(\pi_L d\psi(Z)) \\
&= \pi_L \bar{\partial}_Z(d\psi(Z)) = \pi_L \bar{\partial}_Z(\partial_Z \psi).
\end{aligned}$$

By (13.1) and (13.2) we have

$$\bar{\partial}_Z \partial_Z \psi = \partial_Z \underbrace{\bar{\partial}_Z \psi}_{=0} - \frac{1}{2} S \underbrace{R^{\partial + \bar{\partial}}(Z, JZ)\psi}_{\in L},$$

hence

$$(\bar{\partial}_Z \delta_L)(Z) = 0.$$

Then also

$$(\bar{\partial}_{JZ} \delta_L)(Z) = S(\bar{\partial}_Z \delta_L)(Z) = 0,$$

and therefore $\bar{\partial}\delta_L = 0$.

To prove the holomorphicity of $AQ \in \Gamma(K^2 \operatorname{Hom}(H/L, L))$, we first note that

$$K^2 \operatorname{Hom}(H/L, L) = \operatorname{Hom}_{\mathbb{C}}(TM, \operatorname{Hom}_{\mathbb{C}}(TM, \operatorname{Hom}_+(H/L, L)))$$

carries a natural holomorphic structure. The rest follows from the holomorphicity of A, Q, and the product rule.

Finally we interpret $\delta_{\tilde{L}}$ as a section in $K \operatorname{Hom}_+(\bar{\tilde{L}}, \bar{H}/\tilde{L})$. Note that the holomorphic structure on $\bar{H}$ is given by ∂. From the holomorphicity of A we find, for $\phi \in \Gamma(\tilde{L})$,

$$0 = (\bar{\partial}A)\phi = \bar{\partial}(\underbrace{A\phi}_{=0}) + A\partial\phi.$$

This shows that $\tilde{L}$ is ∂-invariant. Moreover, it is obviously invariant under A and, as a consequence of $AQ = 0$, also under Q. From Lemma 24 it follows that $\tilde{L}$ is invariant under $R^{\partial + \delta}$, and that for a local holomorphic vector field Z and a local holomorphic section ϕ of $\tilde{L}$,

$$\partial_Z \bar{\partial}_Z \phi = \bar{\partial}_Z \underbrace{\partial_Z \phi}_{=0} + \frac{1}{2} \underbrace{SR^{\partial + \delta}(Z, JZ)\psi}_{\in \tilde{L}}.$$

Then

$$
\begin{aligned}
(\bar{\partial}_Z \delta_{\tilde{L}})(Z)\phi &= \partial(\delta_{\tilde{L}}(Z)\phi) - \delta_{\tilde{L}}(\bar{\partial}_Z Z)\phi - \delta_{\tilde{L}}(Z)\partial_Z\phi \\
&= \partial_Z(\delta_{\tilde{L}}(Z)\phi) = \partial_Z(\pi_{\tilde{L}}d\phi(Z)) = \pi_{\tilde{L}}\partial_Z(d\phi(Z)) \\
&= \pi_{\tilde{L}}\partial_Z\bar{\partial}_Z\phi = 0.
\end{aligned}
$$

References

1. Bryant, Robert. *A duality theorem for Willmore surfaces.* J. Differential Geom. 20, 23-53 (1984)
2. Ejiri, Norio. *Willmore Surfaces with a Duality in S^N (1).* Proc. Lond. Math. Soc., III Ser. 57, No.2, 383-416 (1988)
3. Ferus, Dirk. *Conformal Geometry of Surfaces in S^4 and Quaternions.* In: Summer School on Differential Geometry, Coimbra 3/7 September 1999, Proceedings, Ed.: A.M. d'Azevedo Breda et al.
4. Ferus, Dirk; Leschke, Katrin; Pedit, Franz; Pinkall, Ulrich. *Quaternionic Plücker Formula and Dirac Eigenvalue Estimates.* To appear in Inventiones Mathematicae
5. Friedrich, Thomas. *On Superminimal Surfaces.* Archivum math. 33, 41-56 (1997)
6. Hertrich-Jeromin, Udo; Pedit, Franz. *Remarks on Darboux Transforms of Isothermic Surfaces.* Doc. Math. J. 2, 313 - 333 (1997). (www.mathematik.uni-bielefeld.de/documenta/vol-02/vol-02.html)
7. Kulkarni, Ravi; Pinkall, Ulrich (Eds.). *Conformal Geometry.* Vieweg, Braunschweig 1988
8. Montiel, Sebastián. *Spherial Willmore Surfaces in the Four-Sphere.* Preprint 1998
9. Musso, Emilio. *Willmore surfaces in the four-sphere.* Ann. Global Anal. Geom. 13, 21-41 (1995)
10. Pedit, Franz; Pinkall, Ulrich. *Quaternionic analysis on Riemann surfaces and differential geometry.* Doc. Math. J. DMV, Extra Volume ICM 1998, Vol. II, 389-400. (www.mathematik.uni-bielefeld.de/documenta/xvol-icm/05/05.html)
11. Richter, Jörg. *Conformal Maps of a Riemannian Surface into the Space of Quaternions.* Dissertation, Berlin 1997
12. Rigoli, Marco. *The conformal Gauss map of Submanifolds of the Moebius Space.* Ann. Global Anal. Geom 5, No.2, 97-116 (1987)
13. Tenenblatt, Keti. *Transformations of Manifolds and Applications to Differential Equations.* Chapman& Hall/CRC Press 1998

Index

Recent Reprints and New Editions